A Guide to Microsoft Excel 2002 for Scientists and Engineers

A Guide to Microsoft Excel 2002 for Scientists and Engineers

Third Edition

Bernard V. Liengme

St. Francis Xavier University
Nova Scotia, Canada

ELSEVIER
BUTTERWORTH
HEINEMANN

AMSTERDAM BOSTON HEIDELBERG LONDON NEW YORK OXFORD
PARIS SAN DIEGO SAN FRANCISCO SINGAPORE SYDNEY TOKYO

Elsevier Butterworth-Heinemann
Linacre House, Jordan Hill, Oxford OX2 8DP
200 Wheeler Road, Burlington, MA 01803

First published 2000
Third edition 2002
Reprinted 2003

British Library Cataloguing in Publication Data
A catalogue record for this book is available from the British Library

Library of Congress Cataloguing in Publication Data
A catalogue record for this book is available from the Library of Congress

ISBN 0 7506 5613 1

Printed and bound in Great Britain by Martins the Printers, Berwick upon Tweed

Contents

Preface

Microsoft® Excel is a 'number crunching' application with the accountant as its primary target. However, it also provides the scientist or engineer with a very powerful computational tool. True, there are more sophisticated mathematical applications, such as Mathematica, MathCAD, Maple, etc., but none are as widely available as Microsoft Excel. Furthermore, the learning curve for Excel is very gentle; a little learning goes a long way! Once a few basic skills have been mastered, many spreadsheets may be developed in much the same way one would proceed with pencil, paper and calculator, but with more speed, higher precision and greater flexibility.

This *Guide* is designed to give readers a wide range of examples from which they may learn how to apply Excel to problems in their specialized fields. For the student reader, no advanced knowledge of science or engineering is expected and no one who has taken, or is currently taking, an introductory calculus course should find the mathematics difficult. In many cases numerical methods are used to find approximate answers to problems which can be solved by analytical methods. It is a great confidence booster to know you have obtained the correct result and encourages one to try problems for which the exact methods are either very complex or non-existent.

This is very much a practical book designed to show how to get results. The problem sets at the ends of the chapters are part of the learning process and should be attempted. Many of the questions are answered in the last chapter. The *Guide* is suitable for use as either a textbook in a course on scientific computer applications, as a supplementary text in a numerical methods course, or as a self-study book. Professionals may find Excel useful to solve one-off problems rather than writing and debugging a program, or for prototyping and debugging complex programs. A few topics are not covered by the *Guide*: the major ones being the database functions and subroutine modules. These are fully covered in Excel books targeted at the business community and the techniques are applicable to any field.

When I was working on the first edition of this book Microsoft introduced Excel 97. Since then we have had Excel 2000 and then

Excel 2002. While these have only a marginal effect on functionality, there have been some significant changes in the appearance of dialog boxes and in some terminology. This book uses Excel 2002 for its screen captures but readers with earlier versions should have no difficulty following the instructions.

I am indebted to many people: my students and colleagues for helpful suggestions, to readers of earlier versions who e-mailed comments, Nikki Dennis and Matthew Flynn for agreeing to publish the first edition, Rachel Hudson of Butterworth-Heinemann for her invaluable assistance with the present project, and my wife, Pauline, without whom this book would still be a pipe-dream.

I welcome e-mailed comments and corrections (hopefully with a third edition, these will be few and minor!), and will try to respond to them all. Please check my web site and the *Guide*'s web site at www.bh.com/companions/0750656131 for updates.

I hope you enjoy learning to 'excel'.

Bernard V. Liengme
bliengme@stfx.ca
http:/www.stfx.ca/people/bliengme

Conventions used in this book

Information boxes in the left margin are used to convey additional information, tips, shortcuts, etc.

Information box: Boxes like this contain additional information, shortcuts, etc.

Data which the user is expected to type is displayed in a monospaced font. This avoids the problems of using quotes. For example; in cell A1 enter the text `Resistor Codes`.

Non-printing keys are shown as graphics. For example, rather than asking the reader to press the Control and Home keys, we use text such as: press [Ctrl]+[Home]. When two keys are shown separated by +, the user must hold down the first key while tapping the second.

Generally when a new reference is made to a button on a toolbar, a graphic of the icon is shown in the left column. New features in Microsoft Excel 2002 are flagged in the left margin.

 The Save tool

 New to Excel 2002

In the Problems section of each chapter, an asterisk against a problem number indicates that a solution is given at the end of the book. Excel files for answered problems and additional files may be found at: www.bh.com/companions/0750656131.

Website: www.bh.com /companions/0750656131

1 The Microsoft® Excel Window

Concepts

You are probably anxious to start using Microsoft® Excel. However, you must first become familiar with its window and we, the reader and the writer, must agree on some basic terminology. It is useful to conform to the terminology that Microsoft uses so that you will know what topic to search in Help when more assistance is needed. Unless you are new to Windows®, many of the topics covered in this chapter will be familiar. You should learn the names of the various parts of the window and how to access commands using menus and toolbars.

Exercise 1: Anatomy of the Window

Begin this exercise by starting Excel. Your screen will look similar to that in Figure 1.1. There could be some minor differences because Excel allows the user to customize the menus and toolbars. Furthermore, starting with Excel 2000, the items displayed in toolbars and in the initial drop down menus change with usage. A tool that has been recently used will be displayed on the initial menu. We will look at customization in Exercise 5.

It is convenient to divide the window into seven main parts: title bar, menu bar, Standard toolbar, Formatting toolbar, workspace, task pane and status bar. You will be familiar with the first four areas from using other applications so they will be described only briefly.

Title bar
In starting Microsoft® Excel, we have opened a new *workbook*. Because we have not yet saved our work, Excel has given this the default name of *Book1*.

Menu bar
The menu bar provides the user with one way to access the Microsoft Excel *commands*. Commands are actions you perform on your worksheet. Examples are: saving the data to a file, printing a worksheet, changing the appearance of some text, etc.

Toolbars
Toolbars are another, more intuitive and quicker, method of accessing commands. Each *tool* on a toolbar is depicted by an *icon*.

We may speak of, for example, 'clicking on the Copy tool' or the 'Copy button.'

Formula bar

We will examine this more closely in a later chapter. For now, click the mouse in several places within the workspace and watch the information change in the *name box* which is the left-hand part of the formula bar. As you select different cells you will also note how the corresponding column and row headings are highlighted.

Specifications: For other limits (such as the maximum number of characters allowed in a cell) use the Answer Wizard in Help to search for *specifications.*

Workspace

This is the central part of your work. It is here that you will type data and perform calculations. Note how the main part of the space is divided by gridlines into *rows* and *columns.* The smallest unit of space, where a row and a column intersect, is called a *cell.* At the top of the worksheet are the 256 *column headings* starting with A and ending with IV. To the left are the *row headings* numbered 1 to 65,536. How many cells are there on a single worksheet?

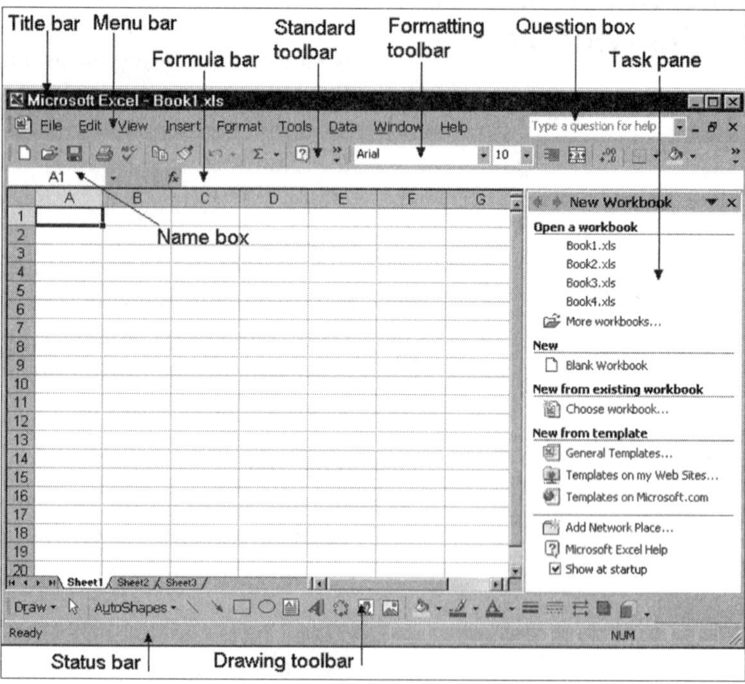

Figure 1.1

What we are looking at in Figure 1.1 is only part of one of the *worksheets* which makes up your *workbook.* To the far right, and at the bottom of the workspace, you will see the vertical and horizontal *scroll bars* which allow you to view other parts of the worksheet. Also at the bottom are the *sheet tabs* which give you access to the other worksheets. Excel normally opens a new workbook with three worksheets. It is possible that your copy of Excel has been configured to give a new workbook some other number of worksheets. We may delete or add extra worksheets to a maximum of 255 depending on the amount of memory in your computer. Later, we will be introduced to chart sheets. We will not investigate *module* sheets which were made redundant with Microsoft Excel 95.

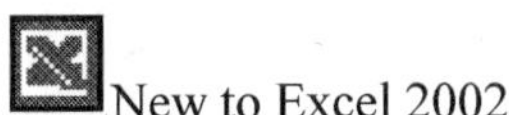
New to Excel 2002

Task pane
In Excel 2002, to the right of the workspace is the task pane. This is designed as a productivity tool. It adds no new functionality but groups frequently used tools in one place. However, it greatly reduces the amount of the workspace that is visible so you may wish to close it. There is more on this topic in Exercises 4 and 5.

Question box
In Excel 2002, to the right of the menu bar is the question box. This provides a handy way of starting Help and recalling earlier topics you have looked up.

Status bar
The status bar provides information. To the left is the *message area*. If your mouse pointer is within the workbook area, this should be showing the word 'Ready'. To the right are some sculptured boxes called the *keyboard indicators*. Press the CapsLock key a few times and watch the text 'CAPS' appear and disappear.

Cells and ranges
Clearly, we need a way to refer to a specific cell on the worksheet. We have seen that a cell occurs at the intersection of a column and a row. We speak of a *cell reference* which is a combination of the column heading and the row number. The cell at the top left, which is at the intersection of column A and row 1, has the cell reference A1. The cell below is A2 while the cell to the right is B1. This method of referring to cells using the column letter is called the A1 method.

There is another method in which the column letter is converted to a number; this is called the R1C1 method since the top left cell has the address R1C1 using this method. We shall not pursue this.

Carefully note that in this paragraph we have not spoken of a cell *name*. In Chapter 2 we find that this has a very specific meaning.

A *range* is a rectangular block of cells. The cells A1, A2, A3, B1, B2 and B3 form a range which we can refer to using A1:B3. In general a range is denoted by the cell references of the top left cell and the bottom right cell separated by a colon. Since a range may be a one by one block, the word *range* may also refer to a single cell. In a later chapter, we will learn how to reference a range from another worksheet in the current or another workbook.

Exercise 2: The Workspace

In this exercise you will learn how to enter data into a cell, edit this data and move around (navigate) the worksheet. We also introduce the *formula* concept. By the end of step (g) of this exercise your worksheet should resemble that in Figure 1.2, complete with misspellings. If you do not have time to complete the exercise in one session go to step (a) of Exercise 3 to see how to save your work.

	A	B	C
1	Area and Perimeter calculations		
2			
3	length	6	
4	width	7	
5	area	42	
6	perimetre	26	

Figure 1.2

(a) We wish to enter some data into cell A1 so this needs to be the *active cell*. There are two ways of knowing which is the active cell. Look in the name box of the formula bar (it is just above the A column heading). Does it say 'A1'? Alternatively, in the worksheet is there a border around cell A1? This is called the *cell selector*. If A1 is not the active cell, the quickest way to make it so is to press Ctrl+Home.

(b) With A1 as the active cell, type the heading Area and Perimeter calculations and then press the Enter key. Note how the active cell becomes A2. Pressing Enter always moves the cell selector down one row. If you make a typing error continue on; we will see how to make corrections later. Press Enter once more to move to A3.

Completing an entry: A cell entry may be completed in one of several ways. They differ in how they relocate the cell pointer.

We may use the keys:
[Enter↵], [⇧Shift]+[Enter↵], [Tab⇆], [⇧Shift]+[Tab⇆], [↑], [↓], [←], [→]

We may also use the Enter button (✓) on the formula bar. This displays only when a cell is being modified.

You are urged not to develop the habit of clicking another cell to complete an entry; this will give erroneous results when you start working with formulas.

If you start an entry in the wrong cell, use the [Esc] key or the Cancel button (X) on the formula bar to terminate the action.

(c) In A3 type the word length but do not press [Enter]. This time use the mouse to click the Enter button (the green ✓ button) in the formula bar – see Figure 1.3. This time the cell selector has not moved. We have now seen two methods of completing a data entry: pressing the [Enter] key and clicking the Enter button on the formula bar.

(d) To move to cell A4 , press the [↓] or the [Enter↵] key. Do not left click the cell A4. It will get you to A4 but this is a bad habit. If you are working with a formula, clicking another cell enters its reference into the formula. Continue typing the words width, area and perimetre in A4, A5 and A6, respectively. We will correct the incorrect spelling later.

(e) In cells B3 and B4 enter the values 6 and 7, respectively.

(f) In B5 we will use a formula to calculate the area. In this cell type =B3*B4. There are two important items to note here: (1) a formula always begins with the equal sign, and (2) the multiplication operator is not × but *. If you pressed [Enter] rather than the Enter (✓) button in the formula bar to complete the formula, return to cell B5 by clicking it once. Notice how the cell now displays the **result** of the formula while in the formula bar we can see the actual **contents** of this cell.

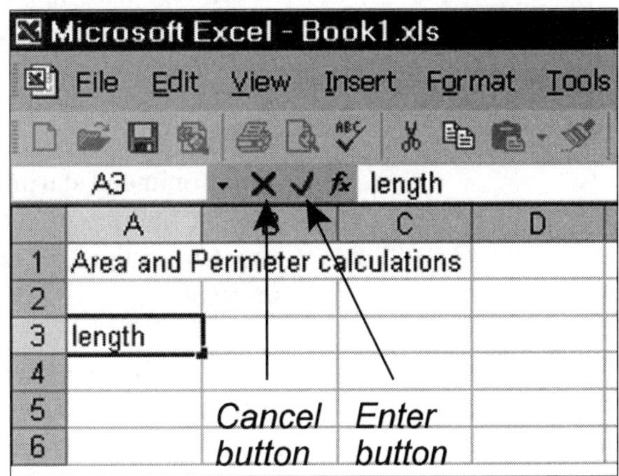

Figure 1.3

(g) In B6 we shall purposely make a mistake. To compute the perimeter, type the formula =2*B3+B4 and click the Enter button (✓). If you pressed [Enter], go back to B6 since we are going to correct it. The formula we need is 2*(B3+B4) to give the correct result; the reason for this is given in Chapter 3. Here we will see one way of making a correction. Move the mouse until the pointer is in the formula bar; note how the pointer changes shape from a hollow + shape to a I (I-beam) shape. Move the I-beam to just after the * symbol and press the (key to insert the opening parenthesis. Now press the End key and add the closing parenthesis,). Click the Enter button (✓) to complete the editing. The cell should now show the correct value.

Your worksheet should now be similar to that in Figure 1.2. Type another number in B3. Note how the values shown in cells B5 and B6 change as soon as the entry is completed.

We have entered three types of data: text, numbers and formulas. Note how text is left justified while numbers are displayed right justified. Later we will see how to change this justification or alignment.

We have learnt how to move around the worksheet using the mouse and the arrow keys. Before completing this exercise, take some time out to explore how the scroll bars work and find what happens when you press PageDown and Page Up. Compare the effects with pressing these keys when the Alt key is held down. Remember that Ctrl+Home will always return you to A1. If your mouse has a wheel between the two buttons you may wish to experiment with it.

With cell A1 as the active cell, see what Ctrl + ↓ does. Try the same command starting first from A3 and then from A6. Return to A1 and find the effect of Ctrl + End. You may also try Go To – either from the Edit menu or by using either F5 or Ctrl+G as a shortcut.

Now we will alter the text in cells A1 to A6. For the purpose of the exercise we will assume that we want each word to start with a capital letter. There are a number of ways of doing this.

(h) Make A3 the active cell and retype the entire word Length to make the correction.

(i) The next correction will be made by editing. Using Ctrl+Home, make A1 active. To edit the cell press F2. The status bar will display Edit rather than Ready and the mouse pointer has become an I-beam. You can also tell you are in Edit mode by

The Undo tool

Shortcut: The keyboard shortcut for Undo is Ctrl+Z. Hold down Ctrl while tapping the Z key.

looking at the lower right corner of the cell selector border. In Ready mode it has a block shape called the *cell handle*; in Edit mode the corner resembles an inverted L shape. Position the pointer before the first letter of 'calculations', press Delete and then an upper case 'C'.

While we are in A1, we can try another experiment. Because A1 contains text and the cells to its right are empty, the text in A1 can overflow into B1 and C1. Click on C1, type some text and press Enter. Now we have 'lost' some characters of the last word in A1. To restore it click the Undo key or use the Undo item on the Edit menu.

(j) Double click on A4 and you will find that this too places you in edit mode. Press the End key. What happens? Now press Home. Type a 'W'. Now we have two 'w's because the default in Excel is *insert mode* – anything we type is inserted into existing text. You can toggle between insert and *overtype* (also called *typeover*) mode with the Insert key. In overtype mode the status bar displays OVR and in the cell one or more characters are highlighted. Experiment with the ←Bksp and Delete keys while editing the cell entry to Width.

The Spelling tool

Shortcut: The keyboard shortcut for Spelling is F7.

(k) In steps (i) and (j) we did the editing within the cell. Make A5 the active cell but this time click the mouse in the formula bar and do the editing there. Press Enter↵ when you have made the change. Do the same with A6 changing only the first letter; do not correct its spelling. After clicking the Entry button (✓), click the Spelling button and let Excel find this and any other spelling errors.

Exercise 3: The Menu Bar

In this exercise we look at the main menu bar and find out how to use some of its various commands.

(a) We are going to save the file containing the work you did in Exercise 2. Click File on the menu bar to pull down its menu. Choose the Save item to bring up the Save dialog which is common to all Windows applications. Select the folder you have chosen to keep your Excel file in. I suggest you name this workbook CHAP1.XLS so that we can refer to it again.

Note how the title bar has changed to reflect the new name. Next time you start Excel, the name of this file will appear at the bottom of the File menu. This gives you a quick way of reopening the file. This feature may not be available if you are on a network.

Keyboard alternative: An alternative way to open a menu is: press [F10], use the [→] and [←] keys to select the main menu item, and press [Enter↵].

(b) Next we will explore the Format menu. You may be wondering why some letters in the names of menu items are underlined. You can open a menu using this letter in association with the [Alt] key in place of clicking the mouse. To open the Format menu, hold down [Alt] and press the [O] key. Note the triangle symbol (▸) to the right of the Columns item. When you click on an item with this symbol a *submenu* appears. Click on Columns and note the ellipses (three dots …) after the Width item. When you click on such an item, in a menu or a submenu, a *dialog box* is opened. Click on Width to open its dialog box. Click the Cancel button since we do not wish to change the column width at this time.

(c) Open the Format menu again. Excel 2000 and 2002 display a drop down menu with a downward pointing arrow at the bottom. To make the menu less cluttered, items that you seldom use are not immediately displayed. The menu will expand if you wait or if you click on the arrow – see Figure 1.4. Most expanded menus are much larger than this example! Use the [Esc] key to close the menu. In Exercise 4 we learn how to customize Excel so that all menu items show immediately.

[⌄] Click to expand menu

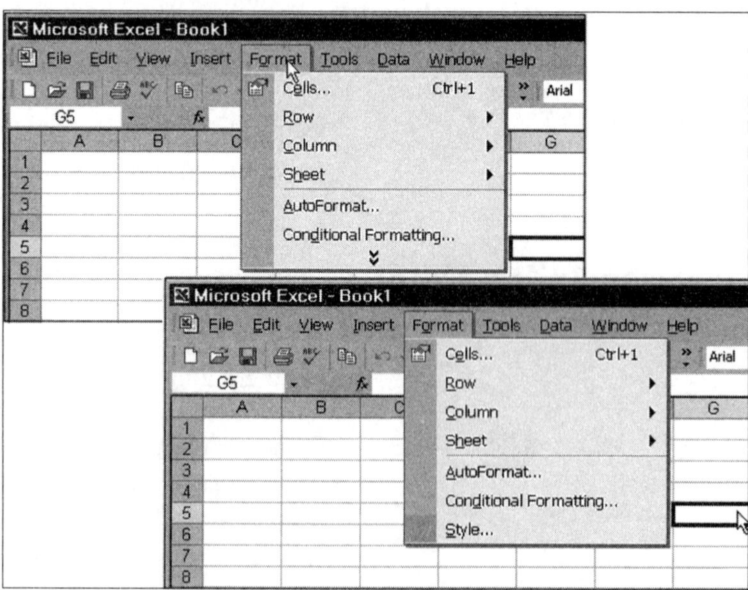

Figure 1.4

(d) You may sometimes wish to use a keyboard shortcut rather than clicking the mouse. Open the Edit menu. If you have not copied something since logging on to your computer, the clipboard will be empty and the Paste feature will be displayed in light text – we say it is *greyed out*. If an item in a menu is greyed out it is currently unavailable to you. To the right of Cut we see the text Ctrl+X, while to the right of Go To is Ctrl+G or F5. These are notes telling you that Cut may be accessed by holding down the Ctrl key while you press the X key without ever opening a menu. Experiment with the Ctrl+C and Ctrl+V to copy the entry in A6 to A10.

To close a menu opened by mistake either click the mouse anywhere in the workbook window or repeatedly press the Esc key until nothing in the menu bar is highlighted.

Exercise 4: The Toolbars

Microsoft Excel provides a number of toolbars. The Standard and the Formatting toolbars are generally displayed all the time. Others appear when the context is right.

The Print tool

An icon on a toolbar allows the user to quickly access one of the Excel commands without the need to open a menu. In addition, since the icons are simple pictures, there is little to learn. For example, if you wish to print a worksheet using the menu bar, you either know the Print command is in the File menu or you must go searching for it. On the other hand, the icon depicting a printer is self-explanatory and easy to find.

The What's This tool

The Spell tool

(a) The pictures for some buttons are less obvious. However, when you move the mouse pointer over an icon without clicking and wait a second or two, Excel displays a *tool tip* naming the tool associated with the icon. For some icons you may find that this method does not give enough information. Let's use the context-sensitive method to get more information. Open the Help menu and click on the item *What's This?* The cursor changes to a question mark. Move the cursor to the *Spell* tool (it has ABC on its icon) and left click it. A larger tool tip with additional information is displayed. Note that if you need additional information about another icon you must visit the Help menu item again.

(b) By default, the Standard and Formatting toolbars are shown side by side in newer versions of Excel to give more room for the worksheet area. Of course, this means that fewer tools can be displayed. Figure 1.5 shows the Formatting toolbar. Note the double arrow on the right of the toolbar; this gives us access to the otherwise hidden tools.

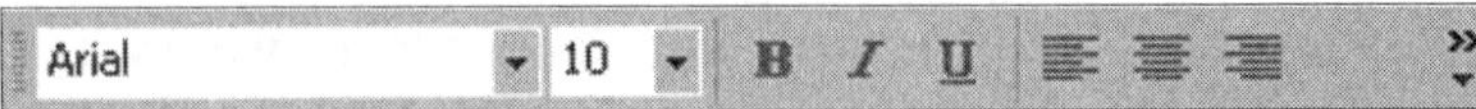

Figure 1.5

(c) The handle on the left of the toolbar may be used to move the toolbar. When the cursor is moved over the handle we get a cursor with four arrows ✥. Carefully drag the Formatting toolbar's handle to the left to reveal more formatting tools but fewer Standard tools. If your toolbar moves below the Standard toolbar, use the menu command Tools|Customize and click on the *Options* tab. Remove the check mark from *Show Standard and Formatting toolbars on two rows* – see Figure 1.6. Note also the options concerning the Menu commands.

If you drag the toolbar down too far it will become a floating menu; it will be a window with its own title bar. To dock the toolbar drag it with the title bar into the area where Excel generally displays toolbars.

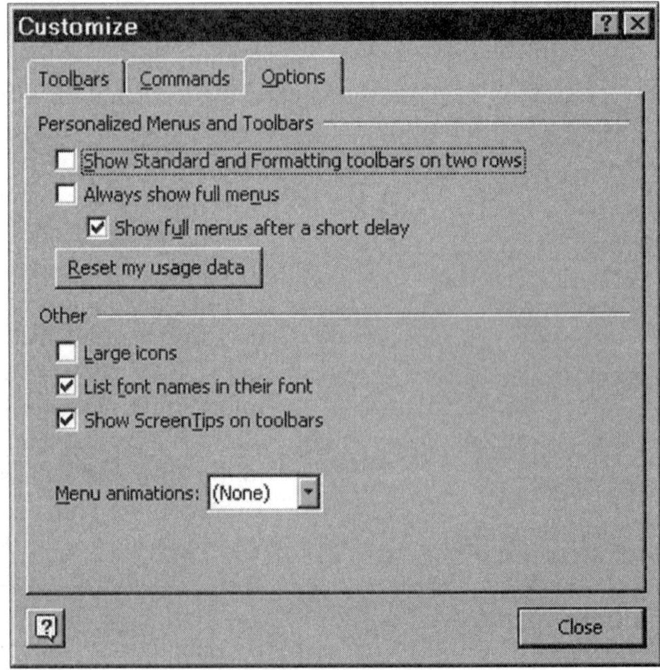

Figure 1.6

(d) While by default Excel displays only the Standard and the Formatting toolbars, it has a wide variety of specialized toolbars. Generally these become visible when needed. So, for example, when we are working on a chart, Excel displays the Chart toolbar provided it is checked in the View|Toolbars menu.

The Drawing toolbar is useful for annotating a worksheet with arrows and other symbols. Use the menu command View|Toolbars to make the *Drawing* toolbar visible. Note that its normal docking place is above the status bar but it can be made floating by dragging its handle. Experiment with the drawing tools and then use View|Toolbars to hide the toolbar.

The command View|Task Pane may be used to open and close the task pane.

Exercise 5: Customizing Menus and Toolbars

The menus and toolbars are set up for the average user. Sometimes we would like something different. Microsoft Excel makes it easy to modify the menus and the toolbars.

(a) A menu or toolbar can be customized only when the Customization window is open. We do this with Tools|Customize. The window is depicted in Figure 1.7. It is similar to Figure 1.6 but this time the *Toolbars* tab is open.

(b) When Figure 1.7 was captured, the Drawing toolbar was displayed. We saw in step (c) of Exercise 4 how to open and close toolbars with View|Toolbars. The Customize window provides another method. Once again all that is needed is to place or remove a check mark in the appropriate box. If you scroll down the list of toolbars you will find another way of displaying the task pane.

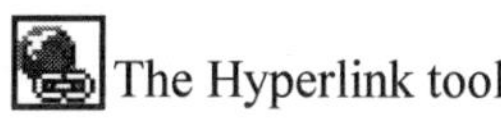

(c) Let us assume the user does not need the *Insert hyperlink* tool. With the Customize window open, place the mouse pointer over the *Insert hyperlink* tool in the Standard toolbar and, holding down the left mouse button, drag the icon off the toolbar. To restore the tool, select *Standard* in the Toolbars tab and click on the Reset button. Click on *Yes* when Excel asks you to confirm the action. The same dragging method may be used to reposition an icon on a toolbar.

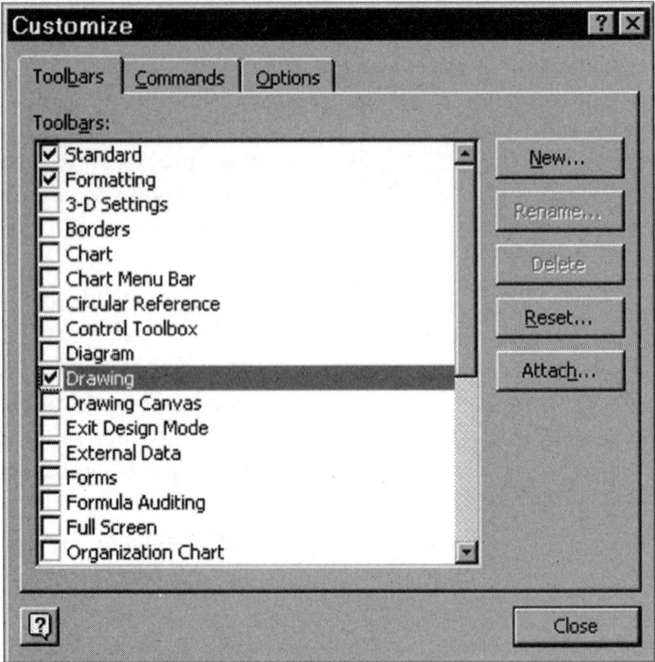

Figure 1.7

(d) To find even more tools, click on the Command tab of the Customize window. Select a topic such as Web and move the vertical slider to review the available tools. To add a tool, drag it onto the toolbar at the required position. Try dragging the *Back* tool on the standard toolbar next to the *Insert hyperlink* tool. Use the Reset button on the Toolbars tab to restore Excel to the default setting.

(e) Items on the menu bar and on popup menus may be dragged to new positions in the same manner as tools. Items may be added to the menu bar. A tool may be placed on the menu bar if you so wish. To restore the menu bar, locate Worksheet Menu Bar in the list of toolbars and click the Reset button. To have all commands display immediately you open a menu bar item, remove the check mark from the *Menus show recently used commands first* box of the Options tab.

Exercise 6: Getting Help

The Help tool

In this exercise we briefly look at how to get help in Microsoft Excel. Suppose we wish to know how to change the width of a column. There is no need to type a complete question such as *How do I change a column width?* You need enter only the key words; so in the case we could use *column width.*

(a) Click on either the Help item of the menu bar or on the Help tool, or press F1.

(b) Unless your copy of Excel has been set to do otherwise (see step (c) below), one of the *Assistants* will appear together with a dialog box into which you type your question. Then click on *Search* and click the item that seems closest to your question – Figure 1.8. This will open the full Help window.

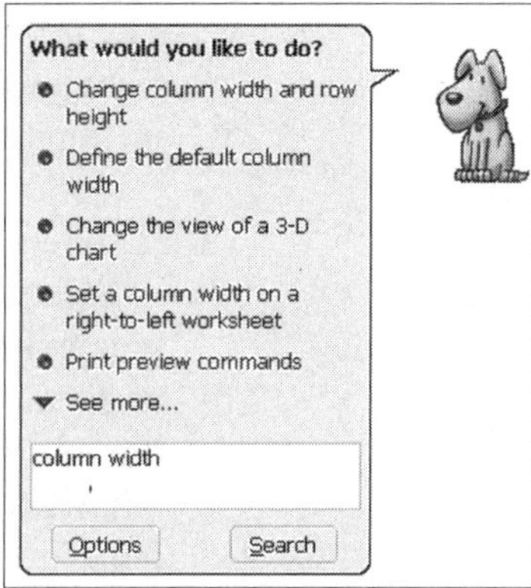

Figure 1.8

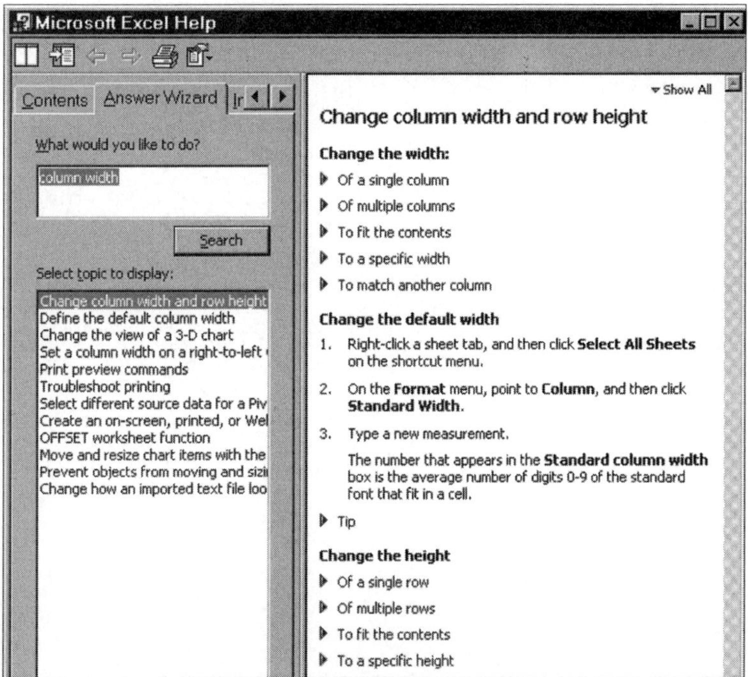

Figure 1.9

If no Assistant appears you will see the full Help window where you can type your question in the text box associated with the *Answer Wizard* – Figure 1.9.

New to Excel 2002

With Excel 2002 one can type the question in the question box – less intrusive than the Assistant. Furthermore, the question box retains your question during a session so you can quickly return to a Help topic.

(c) The information you need is to the right of the Help window. Remember that text in blue will expand to give more information or is a link to another part of Help.

(d) In this step we remove the Assistant. There are two reasons for doing so: (1) some users find the animation annoying and (2) you may wish to see more of the Help system. With the Assistant's dialog box showing (click on the Help tool or on the Assistant graphic) and select *Option* in the box. Alternatively, right click on the Assistant and select the Option item in the popup menu. In the resulting dialog box, remove the check mark from *Use the Office Assistant* box. At this point you may wish to look at the other options.

Restoring the Assistant is simple: click on Help and on the drop down menu select the item *Show the Office Assistant.*

(e) Experiment with the tools on the Help toolbar – you can do no harm! One of the tools hides/unhides one half of the Help window. Did you find it? Open the *Contents* tab and explore that part of Help. The *Index* tab is sometimes useful if the Answer Wizard fails to locate your answer.

Problems

It would take a large book to explain every item on the menu and toolbars. There are two ways to find what various items do: using Help and experimenting. These problems are designed to encourage you to try both ways. You should also learn how to use the Help command Index.

1. Using Exercise 4(b) as a guide, find the purpose of these buttons.

     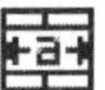 

   Use Help on the menu bar to learn more about these buttons. Using either the Help button or the Help command, find how to hide a column.

2. Construct a worksheet with the misspellings as shown in the figure below.

	A	B	C
1	one		tne
2	two		twenty
3	thre		thirty

   (a) Use the Spelling tool to correct the misspellings.
   (b) Experiment with items on the Format menu to make the text: (i) in A1 italic, (ii) in A2 centred in the cell, and (iii) in A3 centred in the cells A3 and B3.
   (c) Click the Undo button. What happens?
   (d) Locate the Zoom tool (100% ▾) and experiment to find its purpose.
   (e) Find a menu command with the same purpose as the tool just used.
   (f) Locate the Borders tool and put a black border around each cell in A1:A3.
   (g) Select C1:C3 and experiment with the Format|Cells command to find how to draw a red border around the outside of C1:C3.
   (h) Select A1:C3 and experiment with the command Edit|Clear|Formats to see what it does.

+.0 .00 The Decrease Decimals tool

3. On a new worksheet enter 1234567.8 in A1. Observe how the column width is automatically adjusted to accommodate the number. Using the Decrease Decimals tool, format the cell to show no decimal places.

In C1:D1 enter the words: Acceleration, Velocity. The first word is truncated.

Use the mouse to drag the line between the C and D column headers and widen column C to have the whole word displayed.

4. Find out how to use the Office Clipboard. This has the advantage over the Windows Clipboard in that it can hold more than one object. You can even place a series of objects on the Office Clipboard and paste the entire collection with one command.

2
Basic Operations

Concepts

This chapter introduces a number of features needed to prepare a worksheet. We see how to fill a range with a series of numbers, use a simple formula, copy a formula to a range, format numbers and change column widths. The concepts of relative and absolute references are examined. First we review some terminology.

Note: Cell *reference* is the correct term to use when speaking about, for example, A1 or G20. You may find books that use the term cell *address*. However, this is a term that is never used in Microsoft documentation. It was used by Lotus 1-2-3, a spreadsheet application that preceded Microsoft Excel. Surprisingly, Excel does have a worksheet function called ADDRESS that returns a cell reference! Do not use the term *name* in place of *reference*; later in this chapter we will learn that this term has a very specific use in Excel.

Cell reference

This term has been used before; it is how we refer to a specific cell. The reference for a cell consists of the letter for the column in which the cell is located followed by the row holding the cell. The top left cell in a worksheet has the reference A1. To refer to the cell A10 on a worksheet named Data when that is not the current sheet, we would use Data!A10. To refer to a cell in another, open workbook a reference in the form [Book2.xls]Sheet1!A1 is used. A reference to a cell in an unopen file requires the full path and the file name. We may use 'C:\MyData\[Book2.xls]Sheet1'!A1 for example; note that the path and file name are enclosed in single quotes.

Range

A range is a rectangular group of one or more cells. To reference a range we use the top left cell reference separated from the bottom right reference by a colon. Thus A1:B2 refers to the cells A1, A2, B1 and B2. Similarly, A1:A4 refers to the cells A1 to A4. References may be made to ranges in other worksheets and files. We may refer to the range 'C:\MyData\[Book2.xls]Sheet1'!A1:B4 for example.

Active cell

The *active cell* is the one with a heavy black border. A cell must be active if we wish to type data into it. A cell is made active by moving the border with various keys. The key Enter↵ moves it down one row while ⇧Shift+Enter↵ moves it up one row; Tab⇆ moves it right by one column and ⇧Shift+Tab⇆ moves it left one column. The keys ↓↑←→ do the same. Of course, you can move the mouse pointer and click on a cell to make it active. The active cell is surrounded by the *selection box* which has a *fill handle* at the bottom right corner. The reference of the active cell is displayed in the name box of the formula bar. In addition, the column and row headings for the active cell are enhanced.

Selection

If a cell is clicked and the mouse is moved while the button is held down (dragging), a range may be selected. Figure 2.1 shows a selected range. A range may also be selected with ↓←→↑ while holding down Shift. When a range is selected, all the cells other than the active cell have a blue-grey background.

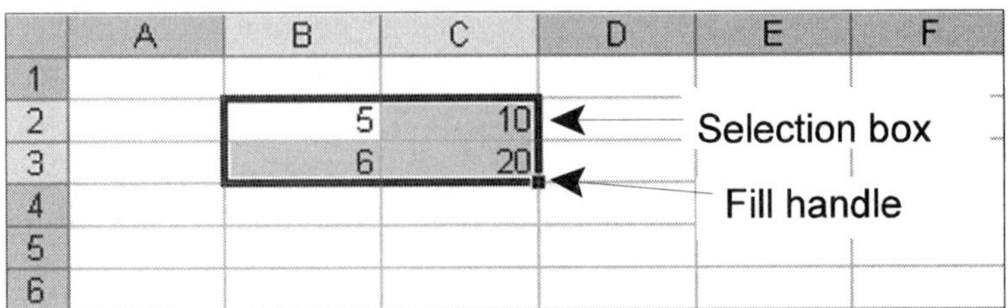

Figure 2.1

Formatting

The manner in which a cell entry is displayed may be changed by *formatting* the cell. Frequently, we use formatting to change the way a cell displays its numeric value. We may change the number of decimals, show the value in scientific notation or with commas, etc. Other formatting options are associated with the typeface and the colour of the displayed entry. The horizontal and vertical positioning of the entry within the cell is also a formatting feature. In addition, cells and ranges may be given borders. In this book we concentrate on formatting for practical rather than presentation purposes.

Note: The format in a cell may be removed by selecting the cell and using the command Edit|Clear|Format.

Exercise 1: Filling in a Series of Numbers

For Exercises 1 to 3, imagine we have a laboratory heating apparatus with a thermometer calibrated in degrees Fahrenheit. We need a table to give approximate Celsius values. On completion of the exercise, the worksheet will resemble that in Figure 2.2.

(a) Open a new workbook. Type the text in cells A1 to B3 as in Figure 2.2. Note that the text is automatically left justified. Remember to press Enter↵ to complete the entry in each cell. You may also complete an entry by any of the methods mentioned in Exercise 2 of Chapter 1 including clicking the green check mark in the formula bar. The check mark is visible only when you are entering or editing an entry.

(b) Select A3:B3 and click the Right Align button on the Formatting toolbar. This will vertical align the text with the numbers we are about to enter.

	A	B	C
1	Temperature Conversion Table		
2			
3	Fahrenheit	Celsius	
4	50	10	
5	100	38	
6	150	66	
7	200	93	
8	250	121	
9	300	149	
10	350	177	
11	400	204	
12	450	232	
13	500	260	

Figure 2.2

(c) Type the numbers 50 and 100 in A4 and A5, respectively.

(d) Rather than type the rest of the values we will use the AutoFill feature. Select A4:A5 and move the mouse pointer over the fill handle. The pointer changes from an open to a solid cross – see Figure 2.3. Drag the fill handle down to A13. Note the screen tip showing the value in the last cell as you drag the mouse down. The range A4:A13 will now have the values shown in Figure 2.2.

The same result can be obtained using the command Edit|Fill|Series. More information on this convenient way of getting data into a worksheet can be found in Help under *Fill in a series of numbers* or *Fill Series command.*

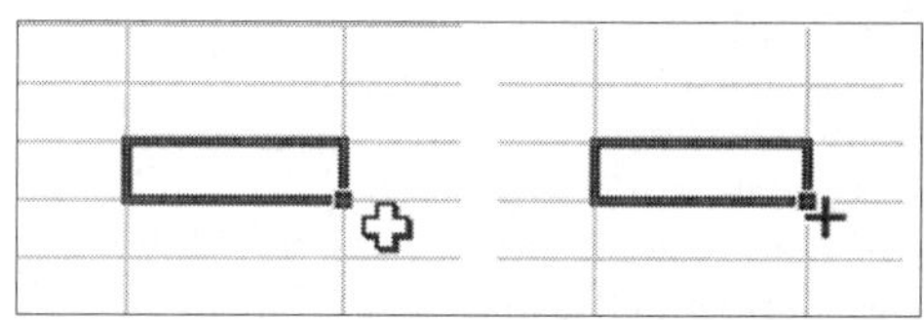

Figure 2.3

(e) Save the workbook as CHAP2.XLS using either the menu command File|Save or the Save tool.

Exercise 2: Entering and Copying a Formula

A formula is an expression telling Excel to perform an operation. Initially, we will limit ourselves to arithmetic operations. All formulas begin with the equal symbol (=). To perform an arithmetic operation we need to follow the = symbol with an arithmetic expression. The expression may contain numeric values, cell references and arithmetic operators. An example of a formula is =A1+A2+A3 which returns the value resulting from the addition of the values in the three cells.

If you were asked to compute the simple expression 3 + 4 × 2 would you reply 11 or 14? Most of us would give the former as would Microsoft Excel. In Excel, and in other programs written in all the major languages, an expression is evaluated following this order of precedence for the arithmetic operators:

Negation (–)
Exponentiation (^)
Multiplication and division (*, /)
Addition and subtraction (+, –)

This precedence order may be overridden by using parentheses. So 2*5 + 6 (or 6 + 2*5) evaluates to 16 while 2*(5+6) evaluates to 22. Note that –3^2 evaluates to 9 not –9 because negation has a higher ranking than exponentiation so it is evaluated as 'the square of negative 3' not as 'the negative of the square of 3'. To get the negative value use –(3^2).

Correcting an error: You are going to make mistakes when learning Excel. Here are hints for quickly correcting them.

1. If you start typing over an existing entry, press Esc to escape before the error is made.

2. You complete an entry and realize you have overwritten an existing one: use the Undo tool, Edit|Undo or the shortcut Ctrl+Z to go back one step.

3. If all else fails, select the cell, press Delete and start again.

(a) On Sheet1 of the CHAP2.XLS workbook, in B4 type the formula to convert Fahrenheit to Celsius, i.e. =(5/9) * (A4 – 32). Do not overlook the equal sign. The spaces around the operators are optional but they do make the formula more readable. If you type lowercase letters for a cell reference, Microsoft Excel automatically converts them to uppercase. The cell displays the value 10. If you make B4 the active cell, you will see =(5/9) * (A4 – 32) in the formula bar.

(b) We need to copy this formula to B5:B12. If you are familiar with Windows you will know how to copy and paste using the Edit menu, shortcuts such as Ctrl+C and Ctrl+V, or the Copy and Paste toolbar buttons. We will use another method. Select B4, 'grab' the fill handle and drag it down to B13 to copy the contents of B4 to these cells below it.

(c) Now we will explore yet another way to copy cells. Begin by selecting B5:B12 and pressing Delete to clear the entries in the range. Place the pointer over the fill handle of B4 and double click. Excel detects that columns A and B form a table and completes the Autofill. This trick works only with vertical tables.

(d) Save the workbook by clicking the appropriate tool.

The values in B4:B13 will be similar to those in Figure 2.2 but will not be integer values. In the next exercise we format the results to show integer values.

Notes on Copying Formulas

Select B5 and look in the formula bar. This cell contains the formula =(5/9)*(A5 – 32) while the formula we copied from B4 has =(5/9)*(A4 – 32). When the formula was copied from B4 to B5, Microsoft Excel automatically adjusted it to reflect its new position, changing the A4 to A5. We need to understand the concept of a *relative reference*. How does Excel interpret A4 in the original formula in cell B4? Rather than 'thinking' that it needs the value in A4, Excel 'reasons' along these lines: the formula is in B4, so A4 refers to the value in a cell one to the left but in the same row. When it copies the formula to B5, the A4 becomes A5 – one cell to the left in the same row as the formula's new location.

Figure 2.4 shows the results of copying a formula to various other cells. Cells A1:B4 contain values. The formula in C2 is =2*A1. Relative to C2, the cell A1 is two columns to the left and one row up. When C2 is copied to C4, the resulting formula is =2*A3 since relative to C4 the cell A3 is two columns to the left and one row up. Similarly, when C2 is copied to D3, we get =2*B2 since, starting at D3, and moving two columns to the left and one row up we arrive at B2.

	A	B	C	D
1	1	10		
2	2	20	2	
3	3	30		40
4	4	40	6	

	C	D
1		
2	=2*A1	
3		=2*B2
4	=2*A3	

Figure 2.4

You should note that if a formula is moved, rather than copied, no adjustment takes place. Later in this chapter we will explore *absolute referencing* which allows us to write a formula in such a way that copying will modify cell references such that only the row, the column or neither is changed.

Exercise 3: Formatting the Results

Our project is almost complete. All that remains is to change the way the results are displayed. For this project it is inappropriate to display so many digits after the decimal; we need integer values.

(a) Select range B4:B13 on the worksheet of the previous exercise. From the Format command on the menu select Cells. In the resulting dialog box select the Number tab – see Figure 2.5. The General format will be highlighted in the Category box. This is the default number format. In most cases, an entry in a cell having the General format is displayed the same way as it was typed. When the cell is not wide enough to show the entire number, the General format rounds numbers with decimals and uses scientific notation for very large and small numbers.

(b) In the Category box, click the Number tab. Change the value in the Decimal places box to 0. You may use the spinner or type the value in the box. Click the OK button to close the dialog box. Your worksheet now displays integer values.

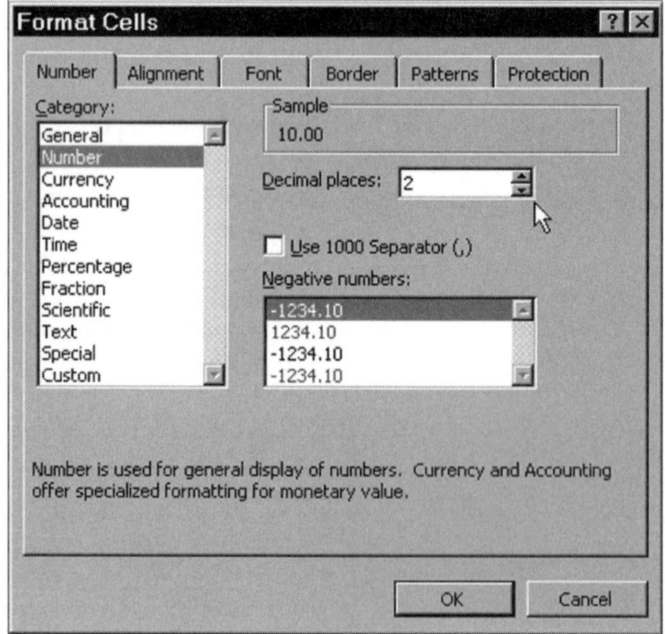

Figure 2.5

(c) There is another way to do this. Click the Undo button on the Standard toolbar to display the original values.

(d) Select B4:B15 and click once on the Increase Decimals button:

+.0 .00 The Increase Decimals tool

.00 +.0 The Decrease Decimals tool

if you try to use the Decrease Decimals tool Excel will make a ping sound to warn you this is impossible – B4 has an integer value so it cannot be made to display any fewer decimal places. All the numbers now display one decimal place. Click the Decrease Decimals button once to display them all as integers.

It is important to know that formatting changes only the way a value is displayed. It does not change the actual value stored in a cell. We look at this in Exercise 4.

(e) Save the workbook CHAP2.XLS.

Notes on Precision and Formatting

Numbers are *stored* with 15-digit precision by Microsoft Excel. The number of digits *displayed* depends on the format and width of the cell. If the user has not applied a format, Excel uses the General format. A cell may be formatted to display a required number of decimal places or to use scientific notation. We do this with the command Format|Cell|Number or by using the Increase and Decrease Decimals button.

In the next exercise we demonstrate that formatting does not alter the stored value. Later we will examine functions that round values to a specified number of decimal places. You should also be aware that when a cell is copied or moved, the target cell gets the same format as the source cell.

Excel can store positive numbers as large as 9.99 999 999 999 999 $\times$ 10^{+307} and as small as 1 $\times$ 10^{-307}. The range for negative values is – 9.99 999 999 999 999 $\times$ 10^{+307} to –1 $\times$ 10^{-307}. The range of values in Microsoft Excel is thus $10^{\pm 308}$ while a typical hand calculator has a range of $10^{\pm 99}$.

You should be aware that conversion from decimal to binary can result in round-off errors. Suppose you perform two complex calculations and expect A99 and B99 to have the same values. Because of round-off errors, the two values may differ by a small amount and the formula =A99 – B99 may not give exactly zero but a value such as 0.000 000 000 000 008 or 8E-15. Just as in decimal notation (base 10) the result for 10/3 cannot be written with infinite precision as a real number, so in binary (base 2) there are some real numbers that cannot be represented exactly.

Enter the values 27.05 and 26.1 in A1 and A2 of an empty worksheet. In A3 enter =A1 – A2 and the value 0.95 is displayed as expected. Now we look at the actual value stored for A3. Progressively increase the number of displayed decimal digits by clicking the appropriate button on the Formatting toolbar. After a while the value 0.94999... is displayed. Clearly there has been round-off error since the result should be 0.95 exactly. Programmers seldom test if two numbers are exactly equal but rather they test if the absolute difference in the two numbers is less than some arbitrarily small quantity.

Exercise 4: Displayed and Stored Values

The purpose of this exercise is to demonstrate that formatting changes only the way in which a value is *displayed*. The *stored* value is unaltered. When completed the worksheet will resemble that in Figure 2.6.

	A	B	C	D	E
1	**Displayed and stored values**				
2					
3	Value	=Value	=Value+2	=2*Value	
4	1.2	1.234	3.234	2.468	
5					
6	The result is different when formatting is done first				
7	1.2	1.2	3.2	2.468	

Figure 2.6

(a) Open the workbook CHAP2.XLS and click on the Sheet2 tab to begin a new worksheet. Begin by typing the text in A1:C3. Entering the text in B3:C3 presents a small problem. The equal sign alerts Excel to use a formula but this is not what we want. Before typing the equal sign, type a single quote (an apostrophe) to tell Excel that you want text.

(b) In row 4 enter the following:
A4: the value 1.234.
B4: the formula =A4
C4: the formula =A4+2
D4: the formula =A4*2

(c) Using the Decimal tools, format A4 to display one decimal place. The cells A4:D4 should show the same values as in Figure 2.6.

(d) Make A4 the active cell. The value displayed in A4 is 1.2 but from the formula bar we see that the stored value is 1.234. Unfortunately, one cannot see the applied format by looking here. You may check how a cell is formatted by selecting the cell and using the Format|Cell|Number command.

If we wish to have a value stored with a set number of decimal digits, we use the ROUND function which is discussed in a later chapter. It is possible to use the Tools|Option command to have Excel use the same precision as the displayed value. This may be useful in financial worksheets but is not recommended.

In the next part of the exercise we see an oddity of Microsoft Excel. When a formula is typed into a cell which has not been previously formatted (i.e. it has the General format) and the formula contains only (i) references to one or more cells with identical formats and (ii) either no operator, or only the addition or subtraction operator, then the cell with the formula gets the format of the referenced cells.

(e) Type the text in A6.

(f) Enter the value 1.234 in A7 and format it to one decimal place.

(g) In B7 enter the formula =A7. The value 1.2 is displayed. Cell B7 has taken on the format of A7 because the formula is a simple reference to a formatted cell. We may format the cell to restore the value of 1.234 if that is required.

(h) In C7 enter =A7+2. Again the value is displayed with one decimal place. However, when =A7*2 is entered in D4 we get 2.468 – this cell does not take on the format of A7. If you use a formula such as =B7+C7 in D7, the result will be displayed with one decimal place since that is how B7 and C7 are now formatted.

(i) Save the workbook.

Exercise 5: Formats Get Copied

In this exercise we see that with the Copy and Paste commands both the values and the formats are copied. However, Excel 2002 provides a way to avoid copying formats. The worksheet will resemble Figure 2.7 when the exercise is complete.

(a) On Sheet3 of CHAP2.XLS, type the text shown in A1:A5. Type 12.555 in B3 and copy this to C3:E3 by dragging the fill handle to the right.

Shortcuts: Windows generally offers more that one way to perform a task. For the Copy and Paste action we could use the menu commands or the tools on the Standard toolbar. But there are two more ways; you may find one of them more convenient.

Keyboard shortcuts: For Copy use Ctrl+C and for Paste use Ctrl+V.

The shortcut menu: Right click on a cell to open the popup menu. It contains both a Copy and a Paste command.

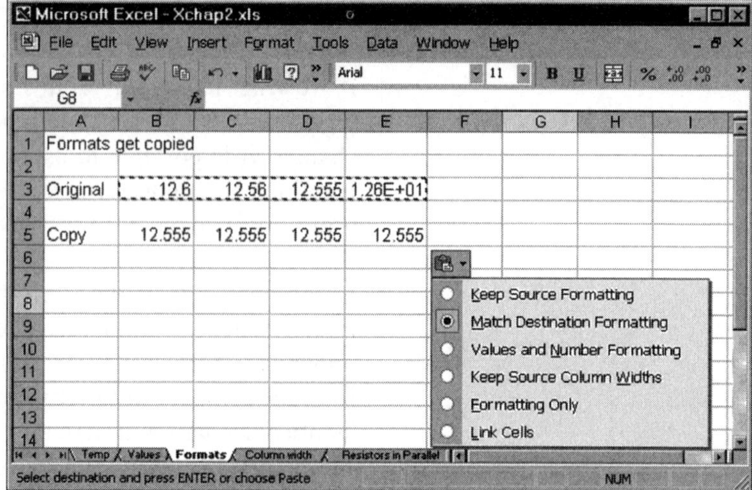

Figure 2.7

(b) Format cells in B3:E3 to display the values as in Figure 2.7.

(c) Select B3:E3 and click the Copy tool. Make B5 the active cell and click the Paste tool. Note that the values in the destination cells are displayed with the same formatting as the source cells.

The Copy operation places material on the Clipboard. To warn you that this has occurred, an animated border (the 'ant track') is placed around the copied material. While the ant track is present you can paste the material on the same worksheet; on another worksheet in the same or another workbook, or in any open Windows document. When you perform any other action in Excel, the material is removed for the Clipboard and the ant track goes way. This is a safety precaution not used in other applications; numeric material erroneously pasted into a worksheet might not be spotted and could lead to incorrect business decisions.

New Excel 2002 feature

(d) Excel 2002 has a new feature associated with the Paste operation: a Paste option smart tag appears. It resembles the Paste icon on the toolbar but when the mouse hovers near it a down arrow is added to the icon. Click the icon to open the smart tag. It offers options to use the formatting from the source or to match the formatting of the destination. Use Help to learn about the other options when you know more about Excel. Save the workbook.

Exercise 6: Too Many Digits

In this exercise we discover what to do when the value in a cell has too many digits to display. There are a variety of ways to change the column width to accommodate the value.

(a) When we opened the worksheet that was to become CHAP2.XLS it very likely had three worksheets since this is the default setting. We need a new worksheet for this exercise. Use the command Insert|Worksheet to make a new one. Now look at the sheet tabs and locate the tab for Sheet4. It was placed to the left of whichever was the current sheet when you use the insert command. Clearly it is in the wrong place. Click on the Sheet4 tab and drag it to the right of the Sheet3 tab to locate it correctly.

(b) Enter the value 123.456 in A1.

(c) Using the Increase Decimal icon, increase the number of decimal digits. After a few clicks the value has filled the cell and Microsoft Excel automatically widens the cell to keep pace with the number of characters displayed. There is, however, a limit; a cell cannot hold more that 256 characters. Furthermore, do not be misled by all those zeros! Microsoft Excel stores numbers to a precision of 15 digits. Adjust the value to about 8 decimal places.

(d) Place the cursor in the column headings and position it on the divider between the A and B headings. The cursor will change to a new shape – see Figure 2.8. Drag the column divider to the left making column A narrower. The cell now displays ########## indicating that the cell is not wide enough to display the value with its current format.

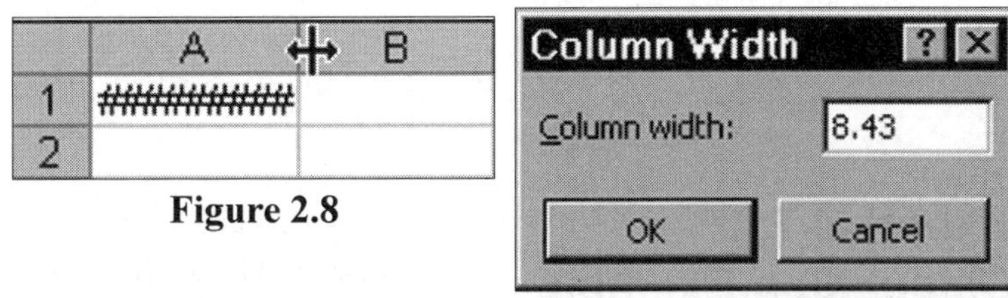

Figure 2.8

Figure 2.9

(e) In the last step we changed the width of a column but we have no idea what the actual width is. This time we will change the width to a specified size. With A1 as the active cell, use the command Format|Column|Width. In the Column Width dialog box (see Figure 2.9) enter the value 8.43. This is the default column width with Arial font of size 10 or 11 points and is large enough to display eight digits.

(f) The cell may still display #######. Use Format|Cell to display the value with 3 decimal places. It should now fit the column.

Shortcut: You can also open the Format dialog box from the popup menu that appears when you right click in a cell.

(g) In A2 enter the value 12345678912. This time Excel does not expand the column to accommodate all the digits because the column has been given a fixed width. Rather, Excel displays the value in scientific format as 1.23E+10 which is to be interpreted as 1.23×10^{10}.

Microsoft Excel behaves differently with text entries. If you type a text entry with more characters than the cell can hold, the entry will overflow into the cells to the right provided they are empty.

(h) Type Sample heading, in B1. Both words are readable but much of the second overflows into column C. Now type Another heading in C1. Most of the characters of the second word in B1 are now lost. B1 is not wide enough to display its contents and text overflow is not permitted now that C1 is occupied.

(i) We now experiment with another way of widening a column. With B1 and C1 selected, use the command Format|Column|AutoFit Selection. The two columns are made exactly wide enough for B1 and C1 to hold their contents.

Shortcut: To change a column width to accommodate the widest entry in that column, double click on the divider to the right of the column header.

The AutoFit command may be used with any type of data, numeric or textual. If we select the column headings rather than a range, the AutoFit command makes the columns the correct width for the cell in each column with the greatest need for space.

(j) Save the workbook.

Exercise 7: Calculation Example

Once a worksheet has been set up to solve a problem, it may be used repeatedly for the same type of problem but with different input values. For example, if you had one quadratic equation to solve, it might not be worth the effort to design a worksheet to do it. If you had a dozen or so equations to solve, then a worksheet solution would be more efficient than using a pocket calculator. Other advantages of the worksheet are (i) the ability to see what values you have used and (ii) the facility to modify the calculation without re-entering all the data.

In this exercise we will design a worksheet to compute the effective resistance of four resistors in parallel. The four resistors (*R1*, *R2*, *R3* and *R4*) in Figure 2.10 have the equivalent resistance value of the single resistor (*Re*) whose value is determined by the relationship shown in the figure.

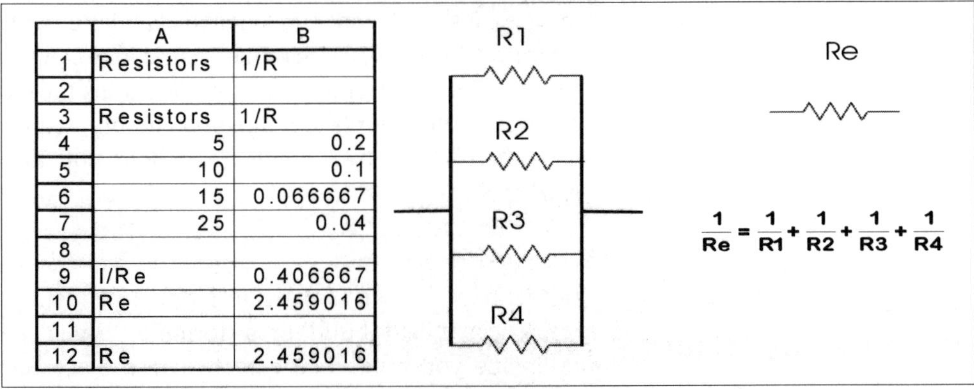

	A	B
1	Resistors	1/R
2		
3	Resistors	1/R
4	5	0.2
5	10	0.1
6	15	0.066667
7	25	0.04
8		
9	I/Re	0.406667
10	Re	2.459016
11		
12	Re	2.459016

Figure 2.10

(a) Using the method in the last exercise, insert Sheet5 in the correct position on the workbook CHAP2.XLS.

(b) Enter only the text and values shown in A1:A10 and B1:B3 of Figure 2.10.

We have started a vertical table in step (b) and completed in Step (c). Excel provides a handy shortcut in such cases. Double clicking the fill handle of B4 causes the formula to be copied down the table.

(c) In B4 enter the formula =1/A4. Copy this formula to B5:B7 by either dragging the fill handle of B4 or double clicking on B4's fill handle.

(d) The formula in B9 is =B4+B5+B6+B7, giving the value 1/*Re*. Remember you may add spaces around the addition operators if you wish. Later we shall use a function to evaluate a summation like this.

(e) The formula in B10 is =1/B9 to give the value of *Re*.

(f) Test your worksheet with the values 2, 2, 4, 4. Since 1/Re will be ½ + ½ + ¼ + ¼ or 1½ , your worksheet should give Re as 0.666667. Whenever possible, check a new worksheet with a few manual (or mental) calculations. Save the workbook.

While our worksheet is able to compute the equivalent resistance of any four resistors in parallel, it cannot be used for fewer. If we enter 0 in A7 (for example), Excel will return the error value #DIV/0! in B7. The same error value will be displayed for all formulas that use B7. Incorporating an IF function (see Chapter 5) in the formulas used in B4:B7 would make the worksheet much more versatile.

What we did in A4:B10 is similar to how we would manually solve this problem with paper, pencil and calculator, writing down every intermediate result rather than using the calculator's memory. Wherever there is repetition (e.g. calculating the reciprocals in this example), the worksheet method simply requires us to copy formulas. These two points may help you design your own worksheets.

Exercise 8: Entering Formulas by Pointing

In this exercise we look at an alternative method to typing cell references when building a formula. When you type a cell reference you must take care to use the correct address. With larger, more complex worksheets, it is easy to make a mistake. The alternative method is to point to the cell with the mouse. It is akin to saying 'use that one'.

For the problem in Exercise 7, clearly we could combine steps (d) and (e) and compute the value of the effective resistence in one formula: = 1/(B4+B5+B6+B7). The parentheses are essential.

(a) In A12 enter the text Re.

(b) In B12, begin the formula by typing =1/(. Now left click on cell B4 and observe the result in the formula bar – the formula is now =1/(B4. Type the plus sign and click on B5. Continue until the formula reads =1/(B4+B5+B6+B7 and click on the green check mark in the formula bar.

(c) If you have entered everything correctly, Microsoft Excel politely points out that there is a small error and offers to correct it by adding the closing parenthesis. Click the Yes button of the dialog box.

(d) Double click on B12 and note the status bar now reads *Edit* rather than *Ready*. But the more obvious change is the *Range*

Finder feature which causes the cells and ranges to which the formula refers to be displayed in colours, and matching colour borders to be applied to the cells and ranges referenced in the formula. This provides a convenient graphical way for us to check formulas. Save the workbook.

Exercise 9: References: Relative, Absolute and Mixed

In Exercise 2 we saw that Excel normally treats cell references as relative references when a formula is copied. There are times when this is not what we need. If the formula refers to the cell A1 we may modify the reference by adding one or more $ symbols.

Reference	Result when formula is copied
=A1	the row and the column may change
=A$1	the row remains constant, the column may change
=$A1	the column remains constant, the row may change
=$A$1	both the row and the column remain constant

A cell reference in the form A1 is called a *relative* reference while $A$1 is called an *absolute* reference. The forms $A1 and A$1 are *mixed* references. Remember that when a formula is copied to the same row, the row reference is unchanged without the need for the $ symbol. Similarly, when a formula is copied to the same column, the column reference is unchanged without the $ symbol.

To demonstrate the use of mixed references, we will develop a simple worksheet that displays a multiplication table as shown in Figure 2.11.

Shortcut: To enter a series of numbers with a constant increment of 1: Type the first number into the cell, hold down Ctrl and drag the fill handle down the column or across the row..

	A	B	C	D	E	F	G	H	I	J
1		2	3	4	5	6	7	8	9	10
2	2	4	6	8	10	12	14	16	18	20
3	3	6	9	12	15	18	21	24	27	30
4	4	8	12	16	20	24	28	32	36	40
5	5	10	15	20	25	30	35	40	45	50
6	6	12	18	24	30	36	42	48	54	60
7	7	14	21	28	35	42	49	56	63	70
8	8	16	24	32	40	48	56	64	72	80
9	9	18	27	36	45	54	63	72	81	90
10	10	20	30	40	50	60	70	80	90	100

Figure 2.11

(a) Insert Sheet6 in the workbook CHAP2.XLS. Start by entering the values 2 and 3 in B1 and C1. Use the Series Fill method to complete the row. Enter the data in column A in a similar manner.

(b) In B2 we need a formula to compute A2 × B1. If we use =A2*B1 we will not be able to copy it. What we need is =$A2*B$1. The $ before the A in the first term ensures that, when the formula is copied across the worksheet, the reference will always be to that column. Similarly, the $ in the second term keeps the reference to row 1 constant when the formula is copied down the worksheet.

The $ signs may be typed as you enter the formula but we shall use another technique. Type =A2 to start the formula. Now press [F4] repeatedly until the formula reads =$A2. Next add *B1 to the formula and again use [F4] to make the formula read =$A2*B$1.

(c) Copy B2 to B2:J10.

(d) Examine the values and the formulas in a few cells to make sure you understand the process.

(e) Save the workbook.

Exercise 10: Editing and Formatting

In this exercise we construct a table to display the pressure of a gas at various temperatures and volumes using the van der Waals equation:

$$P = \frac{RT}{V-b} - \frac{a}{V^2}$$

Note: We could, of course, use the equivalent formula: =(B4*B8)/(A9–E4)–D4/(A9^2).

We would like to be able to change the values of *a* and *b* so that our table may be used with different gases, so we will place these values in their own cells rather than in the formulas. We will also place the value of the gas constant *R* in a cell to provide documentation – and to allow us to change it quickly if we use the wrong value. The final worksheet will resemble that in Figure 2.12. Do not worry about the negative pressure value – the gas has condensed under these conditions and the equation is not really applicable.

(a) Open the workbook CHAP2.XLS and move to Sheet7. Type the text shown in rows 1 to 7 of Figure 2.12. In C4 type CO2. Enter the values in B4:E4. We will format the cells (centring, subscript, etc.) later.

(b) Enter the values in B8:H8 and in A9:A18 using the Series Fill method.

(c) In B9 enter the formula =(B4*B8)/(A9-E4)-D4/(A9*A9). The pointing method from Exercise 8 would be appropriate here. Click the check mark in the formula bar when the formula is complete. The cell should show the value 1374.21. Correct the formula if needed. Examine the formula making sure you understand how it computes the pressure for $V = 0.05$ litres and $T = 250$ K.

	A	B	C	D	E	F	G	H
1				**van der Waals Equation of State**				
2								
3		R	Gas	a	b			
4		0.082058	CO_2	3.59	0.0427			
5								
6				Pressure in atmospheres at varying T and V				
7	Volume			Temperature (Kelvin)				
8		250	260	270	280	290	300	310
9	0.05	1374.21	1486.61	1599.02	1711.43	1823.84	1936.25	2048.65
10	0.10	-0.98	13.34	27.66	41.98	56.30	70.62	84.94
11	0.15	31.63	39.28	46.93	54.58	62.22	69.87	77.52
12	0.20	40.67	45.88	51.10	56.32	61.53	66.75	71.97
13	0.25	41.52	45.48	49.44	53.40	57.35	61.31	65.27
14	0.30	39.84	43.03	46.22	49.41	52.60	55.79	58.98
15	0.35	37.45	40.12	42.79	45.46	48.13	50.80	53.47
16	0.40	34.98	37.27	39.57	41.87	44.16	46.46	48.76
17	0.45	32.64	34.65	36.67	38.68	40.70	42.71	44.73
18	0.50	30.50	32.29	34.09	35.88	37.68	39.47	41.27

Figure 2.12

We need to modify the formula in B9 before copying it to the range B9:H18. There are three considerations: (i) In the formula, B4 refers to the value of the gas constant, D4 to the *a* constant and E4 to the *b* constant. These references must not change when the formula is copied; (ii) On the other hand B8 refers to the temperature and we have a range of these in row 8. When the B9 formula is copied, the reference to B8 must still refer to row 8 but the column must change; we need to replace B8 by B$8; and (iii) The references to A9 must continue to point to the volume values in column A but as the formula is copied the row must change. So we need to use $A9.

In summary, we need to edit the formula to read =($B$4*B$8)/(A$9 - $E$4) - $D$4/($A9*$A9). We shall not retype the formula, rather we shall edit the existing one. We shall do this in a series of steps to illustrate the various options that are available.

(d) Move to cell B9 and enter editing mode by either double clicking or by pressing F2.

(e) We have a variety of options for editing the formula. In cell B9, move the mouse pointer in front of the reference to B4 and click to position the insertion point. Type the $ symbol. Move the insertion point in front of the $ in B4 and type another $ symbol. Now move the insertion point between the B and the 8 in B8 and insert a $ symbol.

(f) Next move the insertion point anywhere within the first reference to A9 – you may have the insertion point in front of the A, between the A and the 9, or just after the 9. Now press F4 repeatedly and watch the reference cycle through the values $A$4, A$4, $A4 and A4. Stop when you have the required value of $A9. Use the same technique to change E4 to $E$4. Click the green arrow in the formula bar to complete the entry and return to ready mode.

(g) We have not completed the editing so we will re-enter the edit mode and explore another way of editing. This time we will activate the edit mode by pressing F2. It has the same effect as double clicking but some users find it more convenient. For variety this time, rather than doing the editing in the cell we will do it in the formula bar. We need to replace D4 by $D$4 and A9*A9 by A$9*A$9. Use whichever method you prefer: typing the $ or using F4. Click on the green arrow in the formula bar when you have completed the task.

(h) B9 should now read: =($B$4*B$8)/($A9–$E$4)–$D$4 /($A9*$A9). Copy it to row 18 and column H. Check that your values agree with those in the table. If they do not, you may need to edit B9 and recopy it.

(i) Now some formatting to improve the appearance of some cells.
   (i) Format the range B9:H18 to display two decimal places.
   (ii) Select A1:H1 and centre the *van der Waals ...* text over these cells using the Merge and Center button. Use the second item in the Formatting toolbar to increase the size of the font to 14. Click the Bold button on the same toolbar.

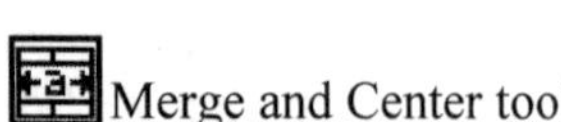
Merge and Center tool

   (iii) Select B3:E4 and centre the entries with the Center button on the Formatting toolbar.

Center Align tool

   (iv) Using the technique in (i), centre the text in A6 over columns A:H and the text in B7 over columns B:H.

(v) In C4 we have CO2 but wish to have CO_2 with the '2' as a subscript. Select C4 and start the edit mode. Highlight the *2* and, from the menu bar, select Format|Cells. Using the Effect portion in the resulting dialog box (lower left), click on the Subscript box to place a ✓ in it. Click the OK button.

(j) We will do more with this worksheet later. Giving the worksheet a name other than Sheet7 will help us locate it. Right click the sheet's tab, select Rename from the menu and type the name VanderWaals. Save the workbook CHAP2.XLS.

Exercise 11: What's in a Name?

In the exercises so far, we have constructed formulas that use other cells' values by using cell references. It is possible to give a cell, or a range of cells, a name. The advantages of this are twofold:(i) it is easier to remember where a value is stored if the cell has a name and (ii) names are always treated as absolute references.

In this exercise we use names with one letter. This is not a requirement, we may use a name such as GasConstant. Note that names are not case sensitive so GASCONSTANT and GasConstant are treated as the same name. When single letters are used, R and C are invalid since Excel uses these letters for other purposes. Similarly, we may not use a name which could be a cell reference. However, we may add an underscore to create names such as R_, c_, X1_, etc.

(a) Open CHAP2.XLS and move to the VanderWaals worksheet. Delete the range B9:H18.

(b) With the range B3:E4 selected use the command Insert|Name|Create. A dialog box similar to Figure 2.13 appears. Excel has detected text in the top row and values in the lower one, so it correctly assumes that you wish to apply the names in the top row to the corresponding cells in the lower row. Click OK.

(c) Move to I1 and use Insert|Name|Paste. In the resulting dialog box (Figure 2.11) click Paste List. This lets us check what names have been assigned to what cells. Note that B4 got the name R_ not R. Delete the list or press the Undo button.

(d) In cell B9 type the formula =(R_*B$8)/($A9 - b) - a/($A9 * $A9). Check its value and copy it to B9:H18.

(e) Save the worksheet.

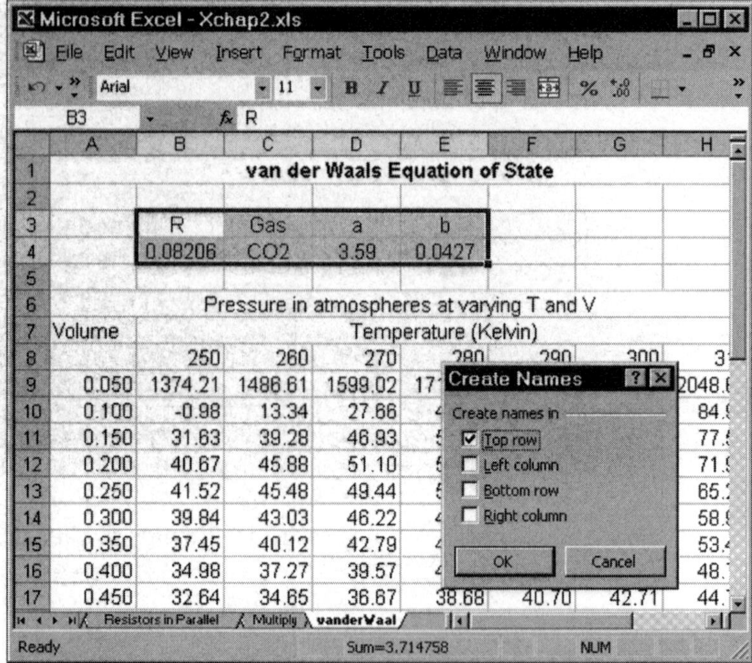

Figure 2.13

We used a semi-automatic method to create names for some cells. We may also select a single cell and use Insert|Name|Define to name a specific cell.

There are a number of ways of referencing a named cell (or range) in a formula. To enter, for example, R_ in a formula we may (i) type the name R_, (ii) point to the corresponding cell with the mouse, and (iii) use the command Insert|Name|Paste, and select the required name and click the OK button or simply double click on the required name.

We began this exercise by deleting the range B9:H18. We did this to show how to build a formula with names. Alternatively, after we had created the cell names, we could have used the command Insert|Name|Apply to replace cell references by cell names.

Shortcut: The Paste Name dialog box is opened with F3.

A named cell or range may be referenced in any sheet of the same workbook. It is possible to use the same name for two cells (or ranges) in different worksheets of the same workbook. Suppose a cell in Sheet1 is named *Mass*, and a cell in Sheet2 has the same name. A reference to *Mass* in Sheet1 will automatically refer to the cell in that worksheet. If, in Sheet2, you need to reference the *Mass* cell of the other sheet, you would use *Sheet1!Mass*. If we move to Sheet3, where no cells are named, what would a reference to *Mass* mean? Generally, it would refer to the cell in Sheet1 since this was first named. However, it would be safer to qualify the name using either *Sheet1!Mass* or *Sheet2!Mass* as required.

Exercise 12: Custom Formats

Engineering notation: The exponent is a multiple of 3.

In Exercise 3 we discovered how to format numbers. We may, for example, arrange to have three decimal places displayed, or to use scientific notation. Excel also permits the user to develop custom formats. To demonstrate this we will format numbers so that the exponent is always a multiple of three – the engineering notation.

	A	B	C
1	Engineering notation		
2	General Format	Scientific Notation	Custom format
3	0	0.00E+00	0.0
4	12.34	1.23E+01	12.34E+0
5	123.45	1.23E+02	123.45E+0
6	1234.56	1.23E+03	1.23E+3
7	12345.67	1.23E+04	12.35E+3
8	1235678.9	1.24E+06	1.24E+6
9	12345678901234.50	1.23E+13	12.35E+12

Figure 2.14

(a) Open CHAP2.XLS and insert a new worksheet. You may need to drag the sheet's tab to the right and correctly locate it.

(b) Enter the text shown in A1:B3 of Figure 2.14. Enter the values shown in A3:A9. When you enter the large number in A9, Excel will display it in scientific notation; change this to the number format with two decimal places.

(c) In B3 enter =A3 and copy this to C3 by dragging the fill handle to the right.

(d) Select B3 and format it to scientific notation with two decimal places. Copy B3 down to B9 by double clicking its fill handle.

(e) Select C3 and open the Format dialog either by using the command Format|Cells or by right clicking the cell and selecting the Format Cells item in the popup menu.

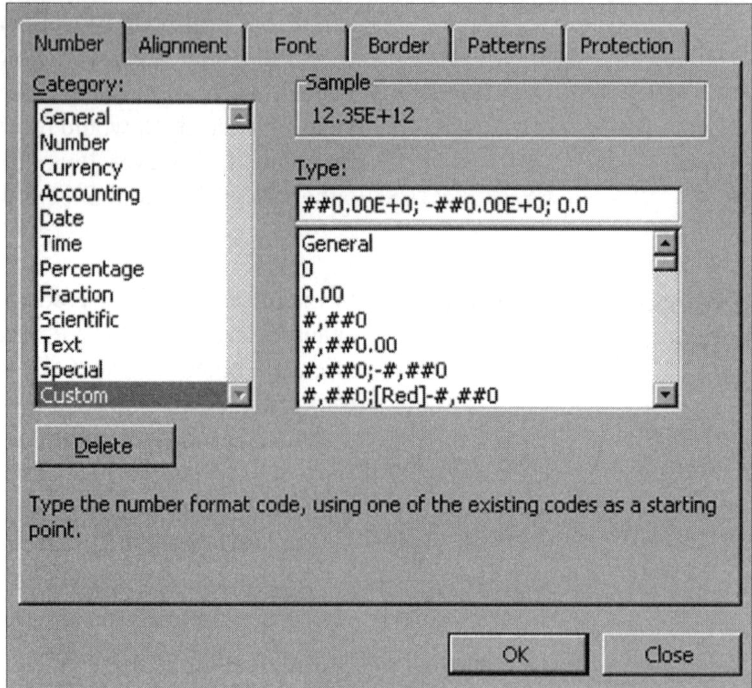

Figure 2.15

(f) Open the Custom tab on the dialog – Figure 2.15. Place the cursor in the Type box and backspace out whatever is there already. Do NOT use the Delete button on the dialog box to do this.

Type in the custom format in this form: ##0.00E+0; –##0.00E+0; 0. A custom format has three parts separated by semicolons. The first part specifies the format for positive values, the second for negative values, and the third for zero. By limiting the number of digits before the decimal to three we have forced Excel to use the engineering notation where the exponent is a multiple of three. Unfortunately, the result is not very pleasing for values less than 1000. We may display as many digits after the decimal as we wish. If you wish not to have zeros displayed use " " for the third part of the format.

(g) Double click on the fill handle of C3 to copy the formula down to C9. Save the workbook.

Exercise 13: Symbols and Such

We may need to use symbols and Greek letters in table headings so we need to know how to get such text as Temp °C and ΔV=πr²Δh into cells. The reader is encouraged to investigate these ways of obtaining such results.

(a) Certain symbols are readily generated by holding down the Alt key and entering a four-digit code on the number pad (it must be the number pad, not the digits on the top row of the 'typewriter' keys). The first two rows of Figure 2.16 show some useful symbols. Thus Alt+0177 gives ±, the plus-minus symbol.

This table is most easily made by entering the numbers in the second row, typing the formula =CHAR(A2) in A3 and copying it across the row. So we may have a cell display °C either by typing Alt+0186 followed by C, or with the formula =CHAR(186)&"C". The former is, of course, more convenient.

Note that we can use this method to get superscripts in text (as, for example, R² and m³) but we can also get these by formatting a normal digit as superscript – see step (i) of Exercise 10.

	A	B	C	D	E	F	G	H	I	J	K	L	M
1	Special characters generated with Alt+0nnn												
2	137	149	150	176	177	178	179	181	186	188	189	190	247
3	‰	•	–	°	±	²	³	µ	º	¼	½	¾	÷
4													
5	The Greek alphabet (using the Symbol font)												
6	a	b	c	d	e	f	g	h	i	j	k	l	m
7	α	β	χ	δ	ε	φ	γ	η	ι	φ	κ	λ	μ
8	A	B	C	D	E	F	G	H	I	J	K	L	M
9	Α	Β	Χ	Δ	Ε	Φ	Γ	Η	Ι	ϑ	Κ	Λ	Μ

Figure 2.16

(b) Greek letters are obtained with the Symbol font. Suppose you want a cell to display ΔV. Begin by entering the two letters DV. Now either select the D in the formula bar, or double click the cell to enter edit mode and select the D in the cell. Next change the font of the selected letter. This may be done with the command Format|Cells and opening the Font tab. But it is more convenient to use the Font box on the Formatting toolbar. Remember that there is no need to scroll down the list, typing

S will jump to the first font beginning with this letter.

The formula bar does not display the formatted value. When you select a cell having some or all the characters formatted in Symbol font, you will see Roman (regular) characters in the formula bar generally in Arial font.

New Excel 2002 feature

(c) If you are familiar with Word you will know that one can use Insert|Symbol to get symbols and Greek characters into a document. Until the latest version, Excel did not support this feature. However, one could use the Windows procedure (Start|Programs|Accessories|Character Map) to copy symbols to the clipboard for pasting into a cell. With Excel 2002 one has access to the Insert|Symbol dialog. Indeed, it is a much improved feature – see Figure 2.17.

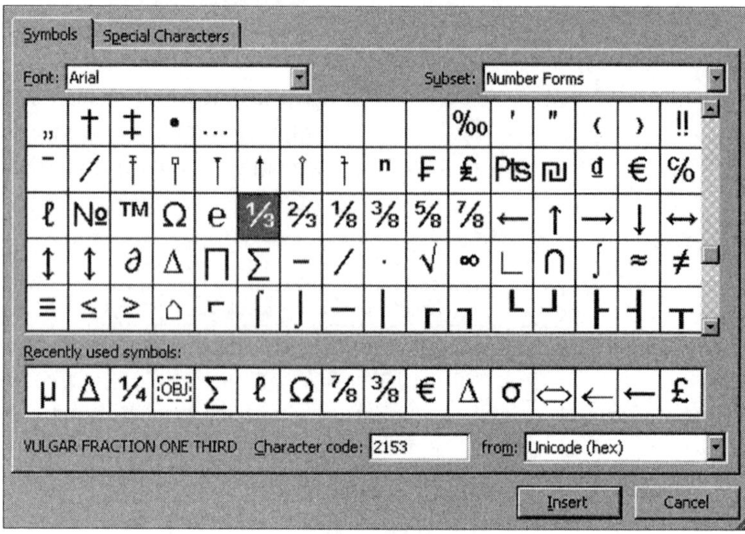

Figure 2.17

Exercise 14: Fractions

From Figures 2.16 and 2.17 we can see that it is possible to enter fraction symbols into cells. It is important to realize that fractions entered in either of these ways are just symbols. They are not numerical values and cannot be used in mathematical operations.

Functional fractions are possible. A cell can display, for example, 2 1/4 but its stored value is, of course, 2.25 – the display results from formatting.

If you mistakenly type 1/8 in A4, Excel will assume you want a data (1 August or 8 January depending on your regional settings). Delete the entry and use Edit|Clear|Format to remove the date formatting.

(a) In Figure 2.18 the entry in A3 was made by typing 2 1/4 – note the space between the integer and the fraction. The entry in A4 was made by typing 0 1/8 – Excel does not display the leading zero. Select each cell in turn and observe the values displayed in the formula bar; they will be decimal values since that is how Excel stores the numbers.

	A	B	C
1	Fractions		
2	N	N × 2	N ÷ 2
3	2 1/2	5	1 1/4
4	1/8	1/4	1/16

Figure 2.18

(b) The formulas in B3 and B4 are =A3*2 and =A3/2, respectively. These are copied down to row 4. It is most likely that C4 will initially display 0. Open the Format dialog and select the Fractions category to adjust this.

Natural Language Formulas

The so-called natural formula feature is mentioned here in case the reader finds a reference to it elsewhere. Your author respectfully advises you not to use this feature.

One of the most hailed new features of Excel 97 was called *natural language formulas*. This allows the use of column or row labels in formulas without creating names. For example, with the worksheet shown in Figure 2.19, the natural language feature allows you, without going through the process of creating names, to use formulas such as =Density*Mass in C2:C3 and =SUM(Mass) in C4.

	A	B	C
1	Density	Volume	Mass
2	5.25	4	21
3	3.5	2	7
4		Total	28

Figure 2.19

The option *Accept labels in formulas* applies to the workbook that was active when the check mark was placed in the option box, and to all new books that are created when the option is in effect.

This feature proved to be a mixed blessing and, while it was retained in later Excel versions, it is switched off when you first install the product. It is controlled by the box labelled *Accept labels in formulas* in the Calculations tab (Tools|Options). The reader is encouraged to leave it switched off and to avoid using this feature since it can easily lead to errors.

Problems

1.* On a new worksheet enter these numbers: 1, 2, 3 and 4 in A1:D1. These represent the values of w, x, y and z, respectively. Do not name the cells for this problem. In row 2 enter formulas to compute the following. Check the results.

(a) $2w - y$

(b) $x^2 + w$

(c) $\dfrac{1}{y^2 - z^2}$

(d) $\dfrac{w + x}{y - z}$

2.* Construct a formula to find: (a) the square root of the value in D1, without using the SQRT function which we look at in the next chapter, (b) the cube root of D2, and (c) the reciprocal of D3.

3. Use the pointing method on the worksheet used for Exercise 7 to build these formulas:
=1/(1/A4+1/A5+1/A6+1/A7) and
=(1/A4+1/A5+1/A6+1/A7)^–1
Do you get the same, correct result?

4. Some of the formatting commands are available on the popup menu that appears when you right click on a cell or a column (or row) heading. Experiment!

5. If P dollars/pounds are invested in a savings account with an interest rate of R per year, compounded M times a year, then at the end of N conversion periods the accumulated amount is given by $A_N = P(1 + R/M)^N$. Construct a table showing the accumulated amount for annual interest rates of 5, 6, 7, 8, 9 and 10% with interest compounding monthly, quarterly and semi-annually. Named cells and formulas with mixed absolute-relative references will work well here.

6. The data validation feature (accessed using Data|Validation from the menu) permits you, for example, to require only integers in the range 1 to 10 in a specified cell. Insert a new worksheet in the CHAP2.XLS workbook and experiment with this feature.

3
Printing a Worksheet

Concepts

Sooner or later you will wish to print your Microsoft Excel worksheet so we will examine this topic now although we have hardly scratched the surface of Excel. In this chapter we will see how to print all the used area or a selected part of a worksheet. We will also explore various options such as changing the header and footers on a printed page, removing the gridlines, having the column and row headings in the printout and making the selected print area fit one page of paper.

Exercise 1: A Quick Way to Print

(a) To begin this exercise we will open the file which was saved in the previous chapter. Open the File menu and look at the bottom. Excel saves the names of the user's last four workbooks. Click on CHAP2.XLS to open your workbook. If you are working on a network, or if you have worked on other files since doing the exercises in Chapter 1, the name of this file may not be present. Use the menu item File|Open to locate the appropriate folder and select CHAP2.XLS. You may wish to type your name in an empty cell such as D4 if you are using a network environment.

(b) Click on the first sheet tab to select the worksheet where you made the temperature conversion table.

The Print tool

(c) Click on the Print button on the Standard toolbar and retrieve your printout. Note that the printed area is A1:D14 since this is the area containing data.

(d) Inspect your worksheet screen. There will be a vertical and a horizontal dotted line. The exact positions depend on the margin setting (we look at these later). Generally, the vertical line runs between columns I and J and the horizontal one between rows 51 and 52. These lines show you what data will fit on a single page in the printout.

Exercise 2: Another Way to Print

This exercise uses the menu to print a worksheet. To demonstrate how this is more versatile than the Print button, we will print just a part of the worksheet.

(a) Select A3:B10 on the fifth sheet in CHAP2.XLS. We know two ways to make this selection – with the mouse or with the combination of the keys [⇧Shift], B, L, R and T. Here is yet another way: click on A3, and while holding down [⇧Shift] click on B10.

(b) Use the menu command File|Print (or the shortcut [Ctrl]+P) to bring up the dialog box as shown in Figure 3.1. In the *Print what* area, click the *Selection* radio button to specify that we wish to print only the selected area.

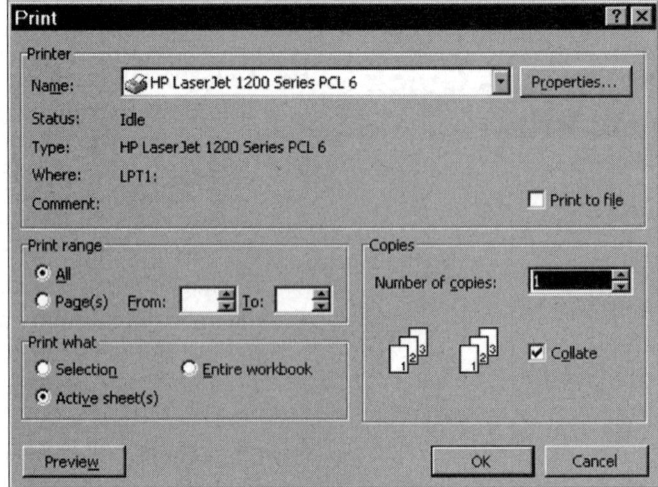

Figure 3.1

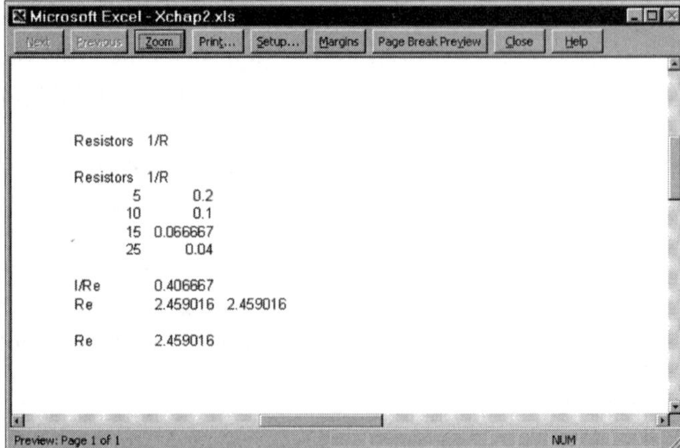

Figure 3.2

(c) To save both time and paper, and to demonstrate another feature, we will not click the *OK* button to start the printing process. Rather, click the *Preview* button. Your screen will display a picture of how the printed page would look – see Figure 3.2. Note that the gridlines of the worksheet will not be printed; we can change this later if needed.

Observe what happens when you repeatedly left click within the Print Preview window: the view is alternately enlarged and reduced. Use the *Close* button to exit Print Preview.

Exercise 3: Page Setup

In this exercise we open the Page Setup dialog box and explore its many options including: scaling the print job to a specified number of pages, setting the margins, requiring that gridlines be printed, adding headers and footers, the orientation of the paper (portrait or landscape), etc.

(a) A worksheet somewhat larger than the ones we have made so far is needed to demonstrate one of the features. Go to Sheet6 of CHAP2.XLS where we made the multiplication table. To extend the table, select A1:J10 and drag the fill handle down to row 60.

The Print Preview tool

(b) Use the command File|Print Preview or click on the Print Preview tool. Experiment with the *Next* and *Previous* buttons in the Print Preview window to see that this worksheet will print on two pages.

(c) Click on the *Setup* button and, if necessary, on the *Page* tab. In the *Scaling* area, click the *Fit to* radio button and leave the values in the two boxes at 1. This will cause Excel to shrink the font size such that your work will fit on one page. The *Fit to* feature does not enlarge, it only shrinks. You can, however, use the *Adjust to* feature to enlarge the print by a specified percentage.

(d) Click on the *Preview* button on the right of the dialog box. You will now find that the worksheet can be printed on one page.

(e) Return to the Page Setup dialog box by clicking on *Setup*. You will see that Microsoft Excel has adjusted the scaling factor to 85%. To return to the original printed size (two pages) you would need to change this back to 100%.

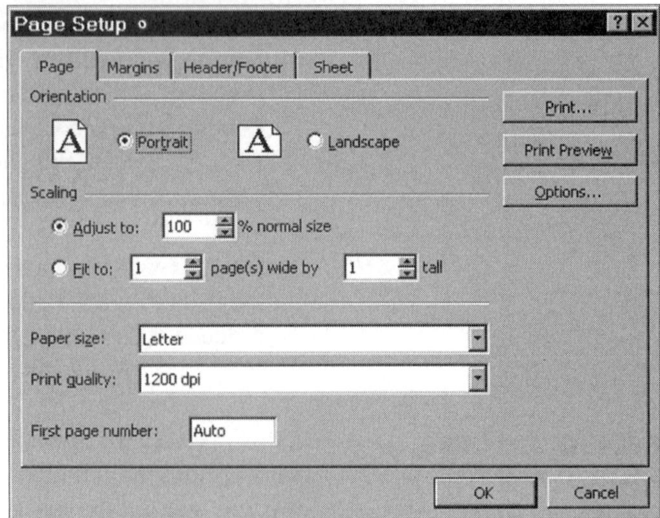

Figure 3.3

(f) Before finishing this exercise you may wish to experiment with the setting to change the orientation from portrait to landscape.

Exercise 4: Changing Margins

(a) With your work showing in Print Preview, experiment with changing the margins by clicking the 'Margins' button. Six dotted lines appear to show the margin positions. The mouse pointer shape changes to a magnifying glass until it crosses one of the margins when it takes on a ✛ shape. When it has this shape, hold down the mouse button and drag one of the margins to a new position.

(b) Click again on the 'Margins' button to remove the margins from the view.

Note that there are six margins: left and right, top and bottom, header and footer. You must be careful not to make the last two so small that the data in the worksheet overlaps the header or footer.

(c) While the method above is useful for a 'quick-and-dirty' fix, it is generally better to set the margins to defined values. Go to the Page Setup dialog box and click the 'Margins' tab.

(d) You can set each margin either by typing a new value in the appropriate box or by clicking the spinners. Set each margin to one inch (or 2.5 cm) and preview your document.

(e) Did you notice the two radio buttons on the Margins tab of the Page Setup dialog box which allow you to centre the worksheet data on the page? Experiment with setting these and preview the result.

Exercise 5: Header and Footer

Headers and footers are for documentation purposes. For course work it is often convenient for the instructor to have the student's name in a header. Many users like to have the file name and the printing date in the printout. As we make use of this feature, remember you can use Preview at any time to see the results.

(a) Access the Page Setup dialog box and activate the Header/Footer tab to obtain a dialog box similar to that in Figure 3.4. The objective is to have your name as the header. If you click the [↓] near the mouse pointer in the figure, you may find your name there. However, if you are using Excel on a network, extra steps are required since the copy of Microsoft Excel is not registered to you.

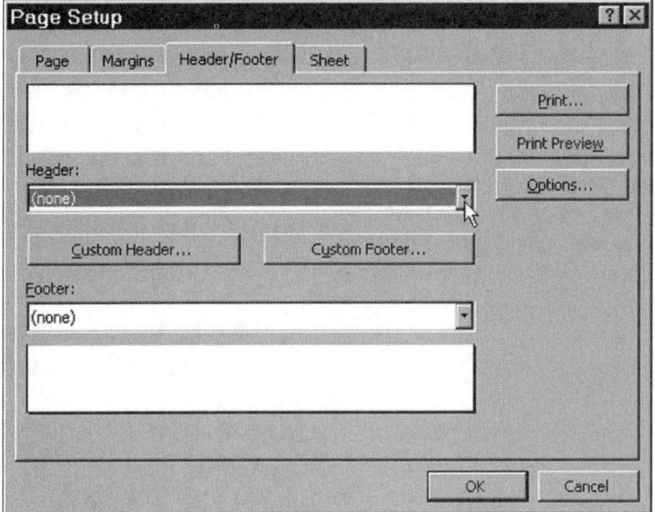

Figure 3.4

(b) Click the Custom Header button to open a dialog box similar to that shown in Figure 3.5. If there is anything in the sections, select and delete it. Type your name in this section and click OK. We will investigate the use of the tools in this window later.

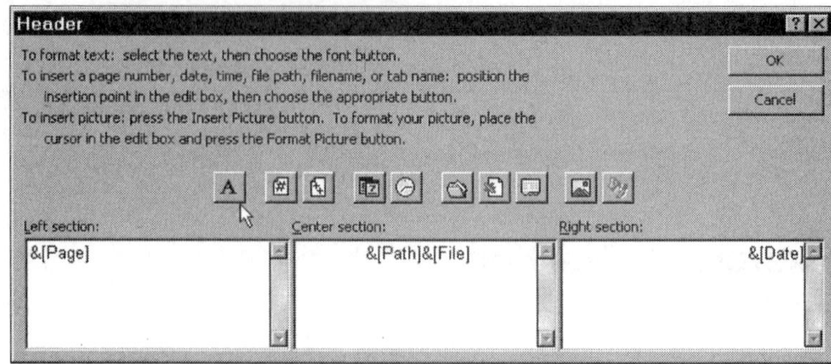

Figure 3.5

(c) Now click the Custom Footer button. The resulting dialog box will be similar to Figure 3.5 except it will have *Footer* as its title. Right click on a button and use the What's This box to discover the purpose of each: reading left to right the tools are:

(i) Font: can be used to change the font of any selected text.
(ii) Page: inserts &[Page] where Page is the page number.
(iii) Pages: inserts &[Pages] to give the total number of pages.
(iv) Date: inserts &[Date] – the date at the time of printing.
(v) Time: inserts &[Time] – the time of printing.
(vi) Path and File: insets &[Path]&[File] – the path and file name (new in Excel 2002).
(vii) File: inserts &[File] – the name of the file.
(viii) Tab: inserts &[Tag] – the name of the sheet.
(ix) Picture: allows a picture to be placed in header or footer.
(x) Picture Format: to resize and otherwise format a picture.

Note: Since the ampersand (&) is used here for coding purposes, to have an ampersand print, you need to type two in a header or footer area.

You may also type in any section. The tools and typed text may be mixed. For example, Page &[Page] of &[Pages] will give text in the form *Page 1 of 6.*

Do not confuse page numbers and sheet numbers. A large worksheet may print as more than one page; it is these page numbers that are being referred to. If you plan to combine pages printed by Excel with those from another application, you may wish to 'customize' the page numbers. The last item on the Page tab of the Page Setup dialog box (see Figure 3.3) is *First page number* and the default value in here is *Auto*. If you wish your first page to be number 12, enter that value in the text box.

(d) Experiment with some of the tools and with adding text. Now preview the worksheet. Return to step (a) if the results are not satisfactory.

Documenting Worksheets

We have seen how to place the date and the time in a footer or header. There may be occasions when you would like the date and/or current time to appear in a worksheet cell. We do not have space here for a complete review of Microsoft Excel's date functions so we will concentrate on one, the NOW() function.

Excel stores dates as serial numbers with the hypothetical date January 0, 1900 as its zero value. If you enter the value 1.5 into a cell and format it as a date, Excel will display 12:00 noon on 1 Jan 1900. Dates can be entered in a variety of formats including 25-May-00 and 25/5/2000. Of course, the order must conform to your Windows Regional Setting. If it is configured for the US format you might enter 05-25-2000.

To display the current date enter =NOW() and format the cell to display the date and time, the date or just the time. Note that the value will change every time the worksheet is recalculated. To enter the current date as a constant use Ctrl + ; and to enter the time as a constant use Ctrl + Shift + :.

A worksheet can also be documented by adding comments to cells with Insert|Comments. A cell with a comment displays a red triangle in the upper right corner. A comment will normally be visible when the mouse lingers over the cell. Alternatively, all comments may be made visible with the command View|Comments.

You may wish to document a worksheet by displaying, as text, a formula next to a cell displaying the value returned by the formula. Click on the cell containing the formula, select the formula in the formula bar using the mouse and click the Copy tool. It is essential you now press Esc. Move to the cell that will contain the documentation, type a single quote (') and click the Paste tool. Remember the Undo tool if you forget the use Esc.

Exercise 6: Gridlines and Row/Column Headings

There are times when a printed worksheet looks better without the gridlines. Sometimes we would like to have a printout in which the row and column headings are displayed; this is useful for documentation and for instructors marking papers. You will find the options for these in the middle section of the Sheet tab of the Page Setup dialog box – see Figure 3.6. You are encouraged to experiment with these and to preview the results.

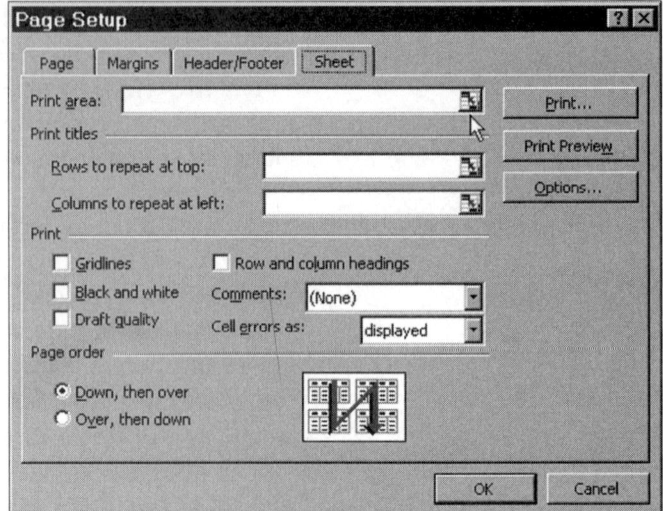

Figure 3.6

Exercise 7: Setting the Print Area

In Exercise 2 we saw one method of printing just part of a worksheet. This method is fine unless you need to print the same area frequently. In which case it is better to specify the Print Area. This can be done in one of two ways.

(a) Go to the sheet in CHAP2.XLS where you completed Exercise 10 of Chapter 2. For the purpose of this exercise we will show how to print just the table of data from this worksheet.

(b) We wish to print A6:H18. We have seen a number of ways of selecting a range. Here is one more. Position the mouse anywhere within the table and press Ctrl+*. From the File menu, go to Print Area and select Set Print Area item. A dotted line will appear around the selected range. Use the Print Preview tool to confirm that this is the part of the worksheet that will be printed.

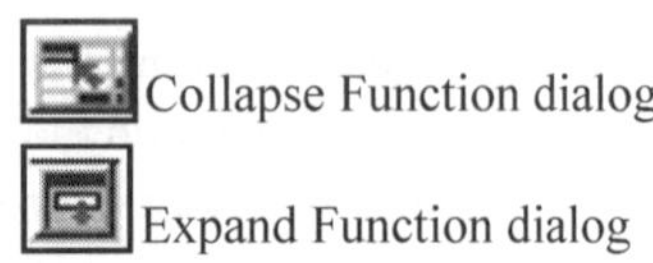

(c) To illustrate the second method of setting the Print Area we need first to clear it with the command File|Print Area|Clear.

(d) Use File|Page Setup and open the Sheet tab (Figure 3.6). At the top of the dialog box is a text box labelled *Print area.* It is possible to type the required range in here. Alternatively, you may use the mouse to select the required range. Click on the red arrow to the right of the Print Area text box to collapse the dialog box. Now you can see your worksheet; only the Print Area part of the dialog box is visible. Select A6:H18 with the mouse and click on the down arrow to the right of the text box to recover the entire Print Setup dialog box.

(e) Click on the Preview button to confirm that you have set a Print Area.

Note: When Page Setup is accessed from Preview, the items for setting the Print Area and the Row and Column Headers are greyed out. To make them accessible you must open the dialog box from the File|Page Setup route.

Exercise 8: Printing Titles

Suppose we have a long table in a worksheet that takes more than one page to print. It is very likely that the first row of the table contains headings (see rows 6 and 7 of Exercise 10 in Chapter 2). Generally it would be convenient to have this information at the top of each printed page. The second item on the Sheet tab allows us to specify which rows and/or columns are to be repeated on each page of output. Extend the table in Sheet7 of CHAP2.XLS down to row 60 and experiment with this feature. To save paper, use Print Preview to check your work.

Exercise 9: Forcing Page Breaks

If you use a word processing application you will be familiar with Ctrl+Enter↵ as a way of forcing a page break. With Microsoft Excel one uses the command Insert|Page Break.

(a) Open Sheet7 of CHAP2.XLS and move to A20. Use the command Insert|Page Break. Note that the page break (shown as a dotted line) is placed between rows 19 and 20; i.e. it is above the cell that was active when the page break was inserted.

(b) With A20 the active cell, use the Insert command again. You will find that Page Break has been replaced by Remove Page Break on the drop down menu. Use this command and the dotted line disappears.

(c) Move to E20 and use the command Insert|Page Break. This time the sheet is slit horizontally and vertically for printing purposes. Of course, you can check this with the Preview tool but, for a change, use the command View|Page Break Preview. If you extended the table in Exercise 9 above, your worksheet will resemble Figure 3.7.

(d) Use the command View|Normal. Make E20 the active cell and remove the page breaks with Insert|Remove Page Break.

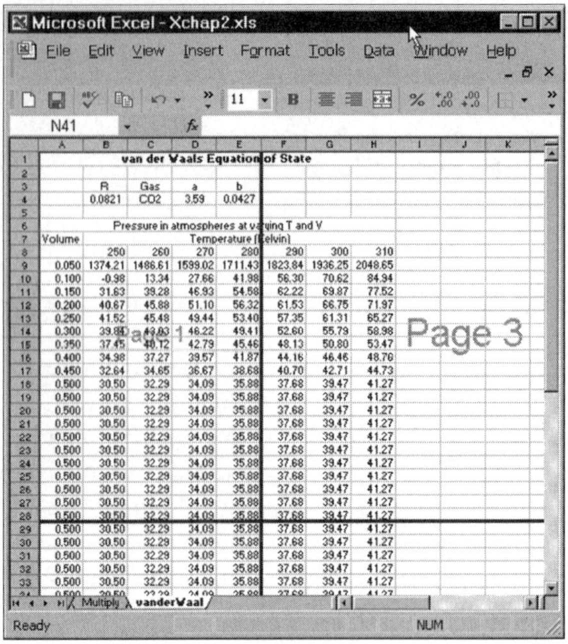

Figure 3.7

Exercise 10: Viewing and Printing Formulas

When a cell in a worksheet contains a formula, its value is displayed on the screen and on the printed page. We can always see the formulas by looking in the formula bar but sometimes we would like a printout showing the formulas for documentation.

	A	B
1	Area and Perimeter ca	
2		
3	Length	6
4	Width	7
5	Area	=B3*B4
6	Perimeter	=2*(B3+B4)

Figure 3.8

(a) Create a simple worksheet with a few formulas – the one created in Exercise 1 of Chapter 1 will do.

(b) Press Ctrl+`. The ` is the key next to the 1 on the top row of the 'typewriter' keys. Your worksheet will resemble Figure 3.8 and will show the formulas.

Note that we can no longer see all the text in A1 since it overflows into B1. If the formulas are long they also may be truncated, in which case select the appropriate column heading and use Format|Columns|AutoFit to have Excel adjust the column widths.

(c) Print or preview the worksheet. Press Ctrl+` to return to the normal view. Readjust the column widths if needed.

Summary of Print Commands

We have seen that many of the commands in the File menu are interrelated in that we can go from one to another. Figure 3.9 shows this graphically.

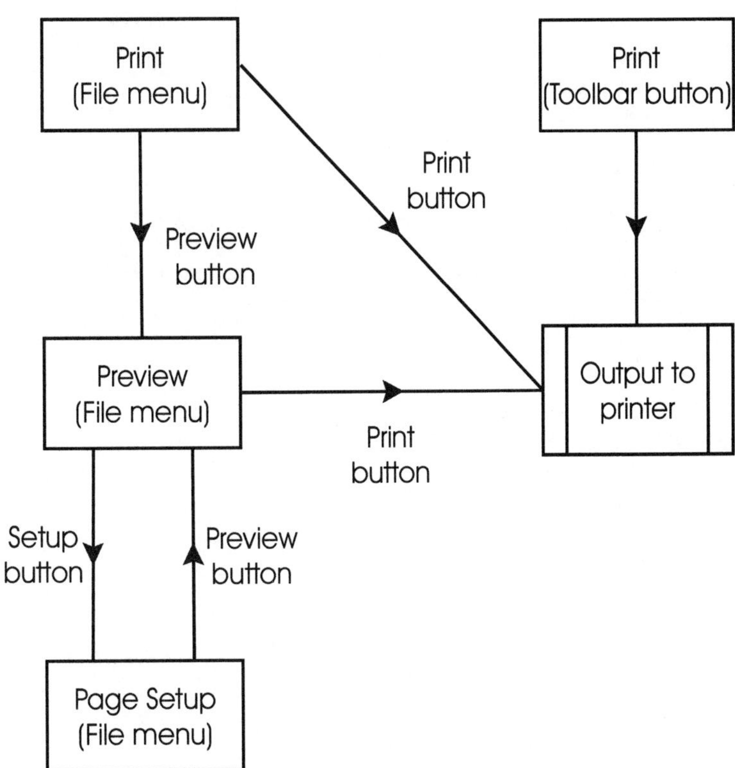

Figure 3.9

4
Using Functions

Concepts

Microsoft Excel provides over 300 worksheet functions which are divided into 10 groups: mathematical and trigonometric, engineering, logical, statistical, date and time, database, financial, informational, lookup and reference, and text. In addition, the user may construct user-defined (custom) functions – see Chapter 8.

Note: The Analysis ToolPak add-in is required for the Engineering functions to be available.

Suppose you wish to know the value of Log(3). We call 3 the *argument* of the function. It is the value that is used by the function to compute the required quantity. Some functions take more than one argument. We say that a function *returns* a value. The *syntax* of a function are the rules for its use.

Figure 4.1 illustrates a formula using a function. The MAX function returns the value of the largest argument. The formula in the figure will return the larger of the value in A1, one of the values in the range B1:B8 or the constant value 10. In this example the arguments are a cell reference, a range reference and a constant.

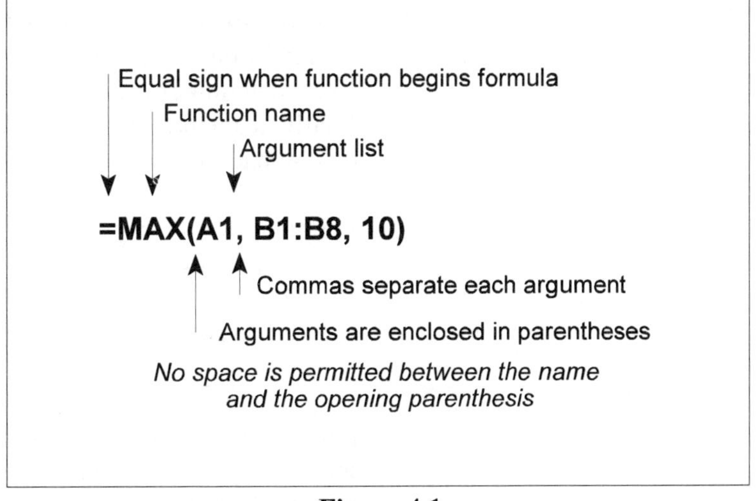

Figure 4.1

Depending on the function, the number of arguments may be fixed, variable or even zero. For example:

zero arguments	=PI()
one argument	=SQRT(A2) or =SQRT(A2/2)
two arguments	=ROUND(A2, 2)
variable number	=SUM(A1:A10) or =SUM(A1:A10,B3,B4)

When the number of permitted arguments is variable, the maximum number is 30 and the number of characters may not exceed 1024. Note that a range such as A1:A100 counts as one argument, not 100.

While some functions require specific types of arguments, most functions permit an argument to be a cell reference, a range reference, a constant, an expression or another function. Certain functions require text type arguments and others require logical arguments. For example:

Cell and range	=SUM(A1, B1:B10)
Named range	=SUM(Xvalues)
Cell and constant	=MAX(A1, 20)
Constant	=LOG10(9.81)
Expression	=LOG10(A1/2)
Function	=SIN(RADIANS(A1))

When a function is used as an argument we use the term *nesting*. Functions may be nested up to seven levels. An example of three-level nesting is =LOG10(MAX (SUM(A1:A4),25)). To interpret this we read from the inside. First the range A1:A4 is summed, then Excel determines the maximum of the sum and the value 25 and, finally, computes the base 10 logarithm of the result of that determination.

Formulas may be constructed from cell references, constants and functions. For example:

=2*PI()	returns 2π
=2.5*SUM(A1:A20)/SQRT(B1)	formula with two functions and a constant

Spaces may be used in a formula to make it more readable. This includes spaces on each side of an arithmetical operator, or on either side of the commas separating arguments in a function call. You may *not* have a space between the function name and the opening parenthesis; you will be rewarded with a #NAME? error.

There is not room in this book to discuss all the worksheet functions. A list of the functions in the various categories can be found by using Help|Contents and then expanding *Function Reference*. You should review the lists before constructing a complex formula or worksheet. For example, suppose A1:A10 contains some numeric values and you wish to find the sum of the squares of these values. You may be tempted to use B1:B10 to hold the squared values and then sum that range. However, Excel provides a function to compute this value; use Help to find its name.

Some functions are described as *array functions* and need to be entered in a special manner. We examine some of these later.

A number of errors can arise with formulas and functions. When this happens, Excel displays one of these error values.

#DIV/0!	Division by zero.
#NAME?	A formula contains an undefined variable or function name, or a space between the name of a function and the opening parenthesis.
#N/A	No value is available.
#NULL!	A result has no value.
#NUM!	Numeric overflow; e.g. a cell with =SQRT(Z1) when Z1 has a negative value
#REF!	Invalid cell reference.
#VALUE!	Invalid argument type; e.g. a cell with =LN(Z1) when Z1 contains text.

When a cell having an error value is referenced in the formula of a second cell, that cell will also have an error value.

An error you are sure to meet once or twice is the *circular reference* error. A formula cannot contain a reference to the cell address of its own location. For example, it would be meaningless to place in A10 the formula =SUM(A1:A10). If you try this, Excel displays an error dialog box with *Cannot resolve circular reference*. If you click OK, the Circular Reference tool appears to help you find the source of the problem. An uncorrected circular reference results in a message in the status bar in the form *Circular: A10* to warn you of the problem. There are some specialized uses for circular references, one of which is shown in a later chapter.

Exercise 1: AutoSum and AutoCalculate

At the completion of the next three exercises, your worksheet should resemble that in Figure 4.2.

	A	B	C	D	E
1	5		Sum	30	
2	10		Average	10	
3	15		Count	3	
4			Min	5	
5			Max	15	
6					

Figure 4.2

(a) Open a new workbook. Enter the values shown in A1:A3 and the text in C1:C3.

Σ AutoSum tool pre-Excel 2002

Σ▾ AutoSum tool in Excel 2002

(b) Select the cell A4 and click the AutoSum button on the Standard toolbar. AutoSum will select the range A1:A3 for its argument. Press Enter↵ to complete the formula. Cell A4 contains =SUM(A1:A3). Microsoft Excel provides this shortcut for the SUM function because many users need to sum a column (or row) of data.

(c) Move the contents of A4 to D1 using the Cut and Paste buttons, or the command on the shortcut menu that appears when you right click a cell.

New Excel 2002 feature

The AutoSum tool has been expanded in Excel 2002 and users of this version may wish to experiment with the additional features as shown in Figure 4.3.

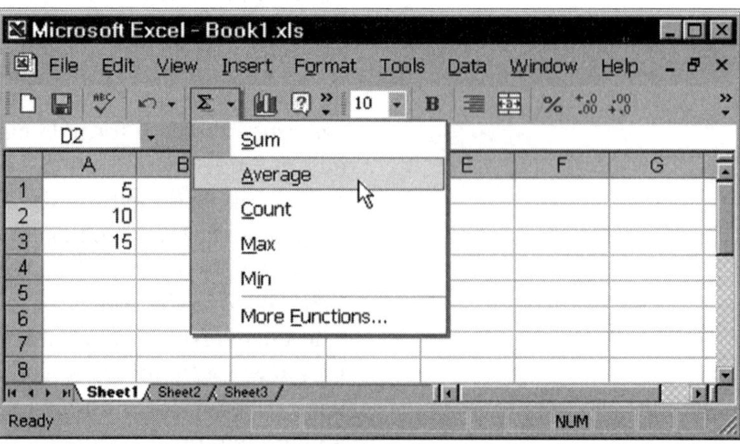

Figure 4.3

(d) Make D2 the active cell and click on the down arrow at the right of the AutoSum tool to reveal a drop down menu. Click on the *Average* item. Excel tries to be helpful and offers to find the average of the range D1 because this is the closest range of numbers. Use the mouse to select A1:A3 and click the green check mark in the formula bar to complete the entry.

(e) You may wish to complete the worksheet by entering the other function from the AutoSum menu. When finished, select D2:D5 and use Delete to clear the cells in readiness for the next Exercise.

For now, ignore the *More Functions* item; it leads to the Insert Function dialog box which we look at in the next exercise.

There may be occasions when you would like to know the sum (or some other statistic) of a range of values but do not need it in the worksheet. The *AutoCalculate* feature was introduced with Excel 97 for this purpose.

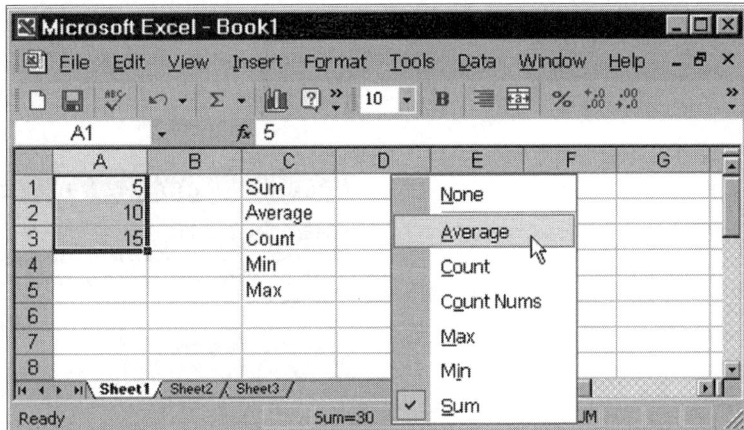

Figure 4.4

(f) Select the range A1:A3 and look at the status bar. In the centre you will see *Sum = 30* – see Figure 4.3. This is the AutoCalculate feature.

(g) Right click anywhere on the status bar to get the popup menus shown in Figure 4.4. This lets you change the statistic reported in AutoCalculate.

(h) Save the workbook as CHAP4.XLS.

Exercise 2: Insert Function

 New Excel 2002 feature

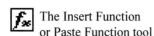

 The Insert Function or Paste Function tool

Excel 2002 introduced some changes in nomenclature. Whereas Excel 97 and Excel 2000 users speak about the *Paste Function* tool and the *Formula Palette*, Excel 2002 users talk of the *Insert Function* tool and dialog box. Also, the location of the tool has been changed. The Paste Function tool is on the Standard toolbar while the Insert Function tool is on the formula bar. Functionally, everything works more or less the same in all Excel versions! This exercise will use the terminology of Excel 2002, other users should readily be able to follow the instructions.

(a) This time we will find the average of the values in A1:A3 of CHAP4.XLS. Select D2 as the active cell. Click the Insert Function button on the formula bar (pre-Excel 2002 users, use the Paste Formula button on the Standard toolbar) to bring up the Insert Function dialog box; see Figure 4.5.

(b) We have no need of the *Search for a function* text box on this occasion since we know the name of the function we wish to use. In the *Function Category* select *All* and under *Function Name* select *AVERAGE*. Later it will be quicker to select a specific category (such as *Statistical*) when you know the function's category. To proceed to the next step, click the OK button or double click the word *AVERAGE*.

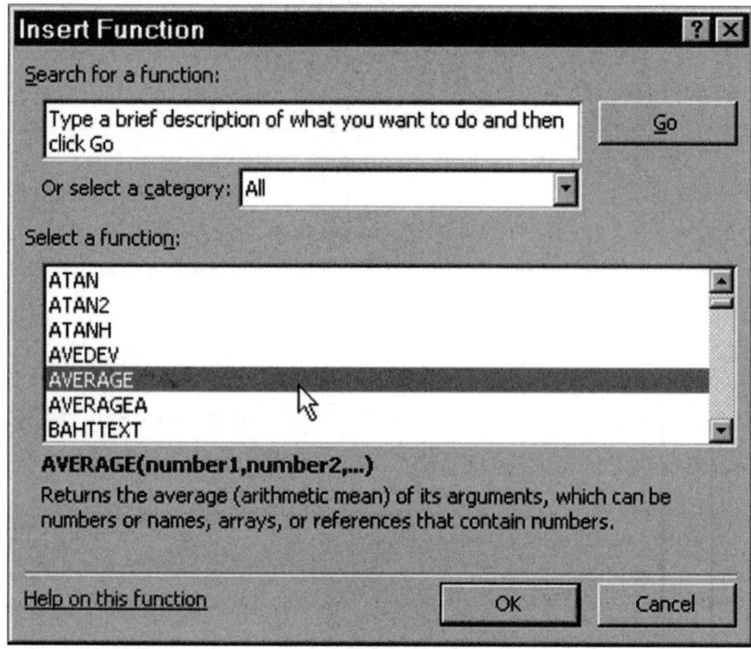

Figure 4.5

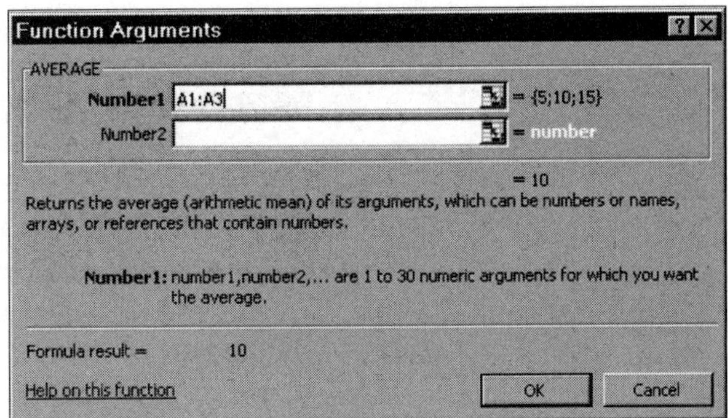

Figure 4.6

Collapse Function box

Expand Function box

(c) The Function Arguments dialog box will appear – Figure 4.6. This gives a brief explanation of the purpose of the function and of each argument. In the first argument box we wish to enter A1:A3. We may do this either by typing or by using the mouse to drag over the range. If the dialog box obscures the required range, click the red arrow at the right of the text box, use the mouse to select the range and click the arrow of the collapsed text box to recover the full dialog box.

The Function Arguments displays the function's value when all the required arguments have been entered. Click the *OK* button to complete the formula. Cell D2 now displays the value 10.

(d) Repeat this process to display in D4 the minimum value of the range A1:A3.

(e) Save the workbook.

What is the purpose of the *Number2* box in the Function Arguments? We may use this to reference other ranges when we wish to find the average of more than one range in a formula. For example, =AVERAGE(A1:A3, A10:A20). A third box (*Number3*) will appear when you do this. Note that the *Number1* argument is shown in bold in the dialog box to indicate it is required while the others are optional.

Exercise 3: Entering a Function Directly

The procedures in Exercise 2 are useful when we are unsure of the function name or the number of arguments it takes. At other times it is simpler to type the formula.

(a) In D5 of Sheet1 of CHAP4.XLS, type =MAX(A1:A3) and press the check mark of the formula bar. Note that had we typed =max(a1:a3), Excel would automatically change the function name and cell addresses to upper case when we completed the formula.

(b) To see another way of entering cell references, delete the contents of D3. Type =MAX(and use the mouse to highlight the range A1:A3. Note that we have 'forgotten' the closing parenthesis. Now click the check mark on the formula bar. Microsoft Excel automatically adds the closing parenthesis.

This is called the AutoCorrect feature. At other times when the correction is not quite so obvious, Excel displays a dialog box with a suggested correction which you must confirm by clicking OK. In other cases, Excel will not be able to make a suggestion and will tell you there is a formula error. Of course, Excel is able to detect only syntax errors not logical errors. If you enter =SUM(A1:A100) mistakenly for SUM(A1:A200), there is no way for Excel to know your intention.

(c) Save the workbook.

Excel 2002 users will have seen a screen tip appear as soon as the opening parenthesis of =MAX(was typed. This is shown in Figure 4.7. Users of earlier versions may obtain similar help using the key combination Ctrl+Shift+A after they have typed the opening parenthesis. The Ctrl+A shortcut to open the Insert Function dialog is explored in Exercise 8.

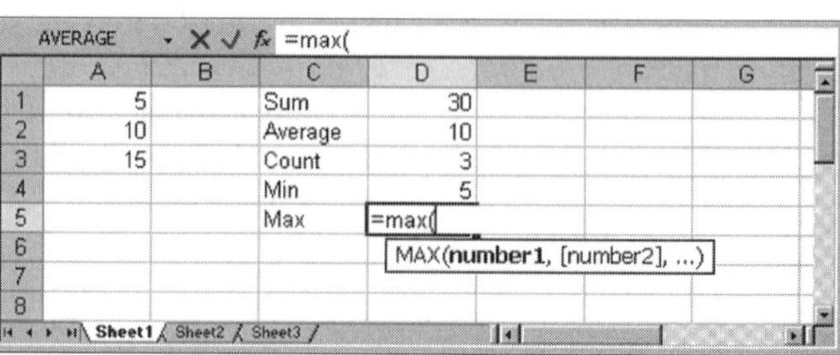

Figure 4.7

Exercise 4: Mixed Numeric and Text Values

Some functions can tolerate arguments referring to cells containing a mixture of numeric and textual values. The functions SUM, AVERAGE and COUNT are amongst these. However, there are some anomalies one should note.

During this exercise, Excel 2002 users will see a green triangle in the top left corner of some cells and, when such a cell is active, an error smart tag (an exclamation mark in a yellow diamond) is displayed near the cell. We explore this topic in the next exercise.

	F	G	H	I	J	K	L
1	Mixing Numbers and Text						
2	1		Sum	4			
3	apple		Average	2		AverageA	1.333333
4	3		Count	2		CountA	3
5			Addition	#VALUE!			
6							
7	Text masquerading as a number						
8	1		Sum	4			
9	2		Average	2		AverageA	1.333333
10	3		Count	2		CountA	3
11			Addition	6			

Figure 4.8

(a) On Sheet1 of CHAP4.XLS enter the text shown in rows 1 to 5 of Figure 4.8 and the values in F2 and F3. The formulas in columns I and L are:
I2: =SUM(F2:F4)
I3: =AVERAGE(F2:F4)
I4: =COUNT(F2:F4)
I5: =F2+F3+F4
L3: =AVERAGEA(F2:F4)
L4: =COUNTA(F2:F4)

Observe SUM, AVERAGE and COUNT simply ignore the textual value in F3. Not all functions are this forgiving. Even the simple formula in I5 cannot cope with this mixture.

For cases when non-numeric values are to be treated as zero, Excel provides the functions AVERAGEA and COUNTA. There is, of course, no need for a SUMA function, since SUM always treats non-numeric data as zero.

(b) Enter the text in F7 and copy F2:G5 to F8. In F9 type '2 .The apostrophe before the digit makes this a textual entry; Excel does not display the apostrophe. Text is normally left aligned but, to make it appear as a number, right align F9. The SUM function again treats the textual value as zero. However, this time a simple addition formula treats the textual value as numeric!

What a headache this worksheet could give to the unwary! A careful worker would find the problem by examining F8:F10 individually while looking in the formula bar. You might wish to experiment with the function ISTEXT and ISNUMBER to find another way to check a column of data such as F8:F10. The rogue value in F9 will also be revealed if you select the column of data and format the cells numeric with two digits.

Exercise 5: Trigonometric Functions

In this exercise we experiment with some of the trigonometric functions which occur in many physical problems. These include: SIN, COS and TAN and their inverses ASIN, ACOS and ATAN. It is important that the user remembers that all computer applications use radians not degrees for angles in trig functions. Since a full circle contains 2π radians and this is equivalent to 360 degrees, the conversion of one representation to another can be made using Radians/2π = Degrees/360. However, it is generally more convenient to use Excel's conversion functions RADIANS and DEGREES.

	A	B	C	D
1	Angle	60	Radians	=RADIANS(B1)
2	Sin	=SIN(RADIANS(B1))		=SIN(D1)
3	Cos	=COS(RADIANS(B1))		=COS(D1)
4				
5	Degrees	=DEGREES(ASIN(B2))	Radians	=ASIN(D2)
6	Radians	=RADIANS(B5)	Degrees	=DEGREES(D5)
7				
8	Side X	1	Side Y	=SQRT(3)
9		=ATAN(D8/B8)		=DEGREES(B9)
10		=ATAN2(B8,D8)		=DEGREES(B10)

Figure 4.9

(a) Open CHAP4.XLS and move to Sheet2. Start a worksheet using Figure 4.9 as a guide; this displays the formulas you should enter. Figure 4.10 shows the expected results.

(b) The formula in D1 converts the degree value in A2 to radians.

	A	B	C	D
1	Angle	60	Radians	1.047198
2	Sin	0.866025		0.866025
3	Cos	0.5		0.5
4				
5	Degrees	60	Radians	1.047198
6	Radians	1.047198	Degrees	60
7				
8	Side X	1	Side Y	1.732051
9		1.047198		60
10		1.047198		60

Figure 4.10

(c) The formulas in B2 and D2 each compute the sine of the angle. In A2 the first thing that is evaluated is the expression RADIANS(B1), then Excel computes the sine of that value. In D2, the argument D1 is already in radians.

(d) The formulas in B3 and D3 similarly return the cosine of the angle.

(e) In rows 5 and 6, we see various ways in which the inverse functions may be used to return the value of the angle in either radians or in degrees.

(f) In B9 and B10 two functions (ATAN and ATAN2) are used to compute the angle given the opposite and adjacent sides. These functions return values in radians. The formulas in D9 and D10 convert the radian values to degrees. Carefully note the differences between ATAN and ATAN2:

ATAN — Uses the form ATAN(opposite / adjacent)
Examples =ATAN(Z2 / Z4) or =ATAN(0.5)
Returns values in the range $-\pi/2$ to $+\pi/2$
Returns an error value if *adjacent* equals 0 since division by zero is undefined.

ATAN2 — Uses ATAN(adjacent, opposite)
Example = ATAN(Y4, Y5)
Returns values in the range $-\pi$ to $+\pi$, excluding π. A positive result is returned for a counterclockwise angle, a negative result for a clockwise angle
Returns an error value if both arguments are zero; 0 when *adjacent* is zero, and $\pi/2$ (90 degrees) when *opposite* is zero.

(g) Verify the remarks about ATAN and ATAN2 by varying the values in B8 and D8.

The functions RADIANS and DEGREES have been used in this example. We could, with less convenience, use the fact that π radians $\equiv$ 180 degrees. For example, the formula in B6 could be replaced by =B5*PI()/180 but there is always the danger of mistakenly inverting the positions of PI() and 180 when using this form.

The hyperbolic functions and inverses (e.g. SINH and ASINH) are also provided in Microsoft Excel.

Another useful function is SQRTPI. For example, SQRTPI(2) returns $\sqrt{2\pi}$ and may be more convenient than =SQRT(2*PI()).

Note: Thanks are due to Chip Pearson for this insight.

You can have Microsoft Excel display your angles in the form 45:30:10, as shown in Figure 4.11. We take advantage of the fact that time and angular measurements have similar formats.

	A	B	C	D
12	Angle	45:30:10	Sin	0.713284
13	Sin	0.505	Angle	30:19:53

Figure 4.11

(h) Enter the text shown in A12, A13, C12 and C13 of Figure 4.11. Enter the value shown in B12. The value in B12 is treated as 45.50278 hours or 1.895949 days, so the formula bar displays 01/01/1900 9:30:10 PM. If you give this cell a general format it will display 1.895949. Use a custom format of [h]:mm:ss to return to the original form.

(i) Suppose we need the sine of this angle. Remembering that the actual value stored is 1.895949, we need to multiply by 24 to convert it to 45.50278. The formula in D12 is =SIN(RADIANS(B12*24)).

(j) Conversely, if we wish to compute the angle having a specific sine value, we can use the approach shown in row 13. The formula in D13 computes the angle from the sine value in B13 using =DEGREES(ASIN(B13))/24. The division by 24 enables us to format this cell with [h]:mm:ss. Clearly, great care must be taken in using these formatted values in further calculations.

Exercise 6: Exponential Functions

On Sheet3 of CHAP4.XLS, design a worksheet to show that:

(a) =EXP(2) returns e^2.

(b) =LN(5) returns the natural logarithm of 5.

(c) =LOG10(5), =LOG(5,10) and =LOG(5) all return the logarithm of 5 to base 10.

(d) LOG(8,2) returns the value 3, which is the logarithm of 8 to base 2.

Use Help to discover why (c) is true; i.e. the behaviour of the LOG function when only one argument is used.

Exercise 7: Rounding Function

In Exercise 4 of Chapter 3 we saw that formatting a cell changes the way a value is displayed but not the stored values. Excel provides a number of functions which either truncate or round a value to a required number of digits or to a multiple of some number. Constant values are used in the examples to facilitate the discussion. Clearly, the function would normally be used with cell addresses or an expression as the first argument. A few of the functions have a second argument. While this may be a constant, a cell address or an expression, it is more usual to use a constant.

ABS	Returns the absolute value. =ABS(−12.55) returns 12.55.
CEILING	Rounds a number up (away from zero) to the nearest multiple of significance – cf FLOOR. =CEILING(1.255, 0.5) returns 1.5.
EVEN	Rounds a number to the nearest even integer. =EVEN(3.25) returns 4.
FLOOR	Rounds a number down (towards zero) to the nearest multiple of significance – cf CEILING. =FLOOR(1.255,0.5) returns 1.0.
INT	Rounds a number down to the nearest integer – cf TRUNC. =INT(−5.6) returns −6.
MROUND	Returns a number rounded to the required multiple. =MROUND(6.89,4) returns 8. This function is only available when the Analysis ToolPak is installed.

ODD	Rounds a number to the nearest odd integer. =ODD(4.25) returns 5.
ROUND	Rounds a number to the required number of places. =ROUND(1.378,1) returns 1.4 (one decimal) =ROUND(123.56,−1) returns 120 (nearest 10) =ROUND(123.56,0) returns 124 (nearest integer)
ROUNDDOWN	Behaves similarly to ROUND but always rounds down.
ROUNDUP	Behaves similarly to ROUND but always rounds up.
TRUNC	Truncates a number to an integer – cf INT. =TRUNC(1.55) returns 1 =TRUNC(−5.6) returns −5 INT and TRUNC differ only when the argument is negative.

On Sheet4 of CHAP4.XLS, construct a worksheet to verify the statements made above. Use Insert|Worksheet if needed and drag the tab to the correct place. Figure 4.12 shows how to start it. You may wish to use Insert|Name to give B1 the name *x*.

	A	B
1	Value of x	1.255
2	CEILING(x, 0.5)	1.5
3	EVEN(x)	2
4	FLOOR(x, 0.5)	1

Figure 4.12

Most of us round 4.3 to 4 and 4.6 to 5. But what about 4.5? While many would reply 5, others use the round-to-even rule. Thus 4.5 rounds to 4 as does 3.5. Unfortunately, Excel does not provide a function that follows this rule but one can construct a user-defined function (see Chapter 8) that does.

Note: Credit for this formula goes to John Walkenbach.

There is a very useful formula to round a number to *n* significant digits. You may wish to experiment with =ROUND(A1, A2 −1 − INT(LOG10(ABS(A1)))) where A1 holds the value to be rounded and A2 the number of significant digits required. Note that the number may be displayed with extra trailing zeros that are not to be counted as significant.

Exercise 8: Array Functions

When you use Help to get information about some functions you may be told that they are *array functions*. Array functions generally return more than one value. There are some important things to remember when constructing a formula using an array function:

1. When the formula generates more than one value, you must select the output range before typing the formula.

2. Once the formula is typed you complete it not with a simple R but with [Ctrl]+[⇧Shift]+[Enter↵]. When you do this, Excel encloses the formula within braces { }. You do not type these braces.

3. If you need to edit an array formula you must select the entire range of output values, edit the first entry and complete the edit with [Ctrl]+[⇧Shift]+[Enter↵].

In this exercise we will multiply two matrices with the MMULT function to demonstrate an array function. Do not be concerned if you are unfamiliar with linear algebra. We shall also see a shortcut method to open the Insert Function dialog.

(a) Use Insert|Worksheet to add Sheet5 and drag the Sheet tab to its correct place. Enter the labels shown in A1, A3, E3 and I3 as shown in Figure 4.13. The label in A3 was centred across the three cells by selecting A3:C3 and clicking the *Merge and Center* tool. Enter the values shown in A3:G6.

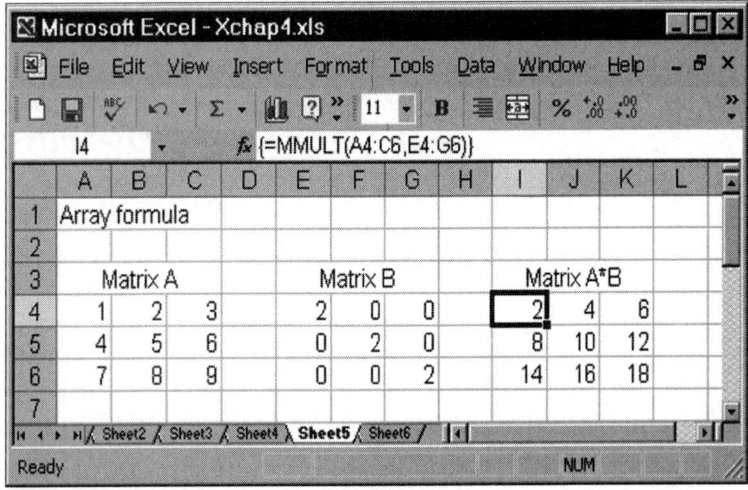

Figure 4.13

(b) Select I3:K6 and type =MMULT. Use Ctrl+A to bring up the Insert Function dialog. Enter A4:C6 as the first argument and E4:G6 as the second. Complete the entry with Ctrl+Shift+Enter.

(c) Observe that with I4 as the active cell, the formula bar shows {=MMULT(A4:C6,E4:G6)}. Microsoft Excel has added the braces, { }, to indicate an array function; the user should never type these braces.

(d) Save the workbook.

Some Other Mathematical Functions

The table below lists some other useful functions which are not covered by the exercises. The examples are shown with constant arguments only for clarity but they are more generally used with cell references or expressions. Functions marked with ‡ are available only when the Analysis ToolPak is installed.

SQRT	Returns the square root of a value; the argument must be positive. =SQRT(9) returns 3.
GCD‡	Returns the greatest common divisor. =GCD(9, 18, 24) returns 3.
LCM‡	Returns the largest common multiple. =LCM(9, 18, 24) returns 72.
QUOTIENT‡	Returns the integer portion of a division. QUOTIENT(28, 9) returns 3.
FACT	Returns the factorial of a number. FACT(4) returns 24.
RAND	Returns a random number between 0 and 1. The formula =RAND()*(b-a)+a returns a value in the range *a* to *b*. The value returned by =RAND() will change every time the worksheet is recalculated. To avoid this and have a random value rather than a formula inserted, type =RAND() then press F9 not Enter.
RANDBETWEEN‡	=RANDBETWEEN(a,b) is equivalent to RAND()*(b-a)+a.

MDETERM MINVERSE MMULT	These return the matrix determinant of an array, the inverse of a matrix, and the product of two matrices, respectively. These are used in Chapter 10.

Working with Time

At the end of the last chapter brief mention was made of how Microsoft Excel deals with dates and time. You may wish to have a time value (for example, the value 85) displayed in the form 1:25 to denote 1 hour and 25 minutes or 1 minute and 25 seconds. To achieve this you must first convert the value to a fraction of a day and then give it a custom format.

This is demonstrated in Figure 4.14. Row 1 contains a series of values. The formula in C2 is =C1–B1 and is copied to J1. The formula in B3 converts the number of minutes in B1 to days with =B1/(24*60). This also is copied to column J after giving it the custom format h:mm. C2 computed a time difference with =C3 – B3 and is similarly formatted. The equivalent quantity is computed in row 5. C5 contains the formula =(C3 – B3)*(24*60) where the factor 24*60 converts days back to minutes.

	A	B	C	D	E	F	G	H	I	J
1	mins	10	25	30	45	50	60	90	120	200
2	diff		15	5	15	5	10	30	30	80
3	h:mm	0:10	0:25	0:30	0:45	0:50	1:00	1:30	2:00	3:20
4	diff		0:15	0:05	0:15	0:05	0:10	0:30	0:30	1:20
5	diff		15	5	15	5	10	30	30	80

Figure 4.14

Problems

1.* You are constructing a worksheet to compute the number of rolls of wallpaper needed to cover a wall. You have a cell called *Length* and another called *Height*. Give the formula needed to compute the number of rolls to be purchased assuming one roll covers 2.25 m^2.

2. The values in the range A3:A53 of your worksheet are between 0 and 100. You need to know how many of these cells have values of at least 50. Use Help or the Function Wizard and find how to use the COUNTIF function.

3.* You have selected the cell F3 and entered the formula =MINVERSE(A3:D6) to get the inverse of the matrix stored in A3:D6. However, you get only one number. What went wrong?

4.* With two cells named *hypot* and *opp* holding values representing two sides of a triangle what formula will return the value of the angle in degrees?

5.* The number of ways of permutating *n* distinct objects taken *r* at a time is given by

$$P = \frac{n!}{(n-r)!}$$

Assuming that *n* is stored in A10 and *r* in B10, give the formula required to compute *P*. You will find one method in this chapter; the other may be found using Help.

5
Decision Functions

Concepts

The functions introduced in this chapter are useful when making decisions. They include the IF function, the logical functions AND, OR and NOT which enable one to make compound tests, and functions such as VLOOKUP, INDEX and MATCH that look up values from tables in the worksheet. We shall also explore the use of SUMIF and COUNTIF. Some array formulas are explored.

The IF and the Logical Functions

The IF function is used when you want a formula to return different results depending on the value of a *condition.* As a simple example, suppose A2:A21 contains the grades of 20 students and you wish to have the word 'Pass' or 'Fail' in the B column depending on whether the student's grade is 50 or greater. The formula =IF(A2>=50, "Pass", "Fail") is typed into B2 and copied to B2:B21. Figure 5.1 shows the syntax for an IF function formula.

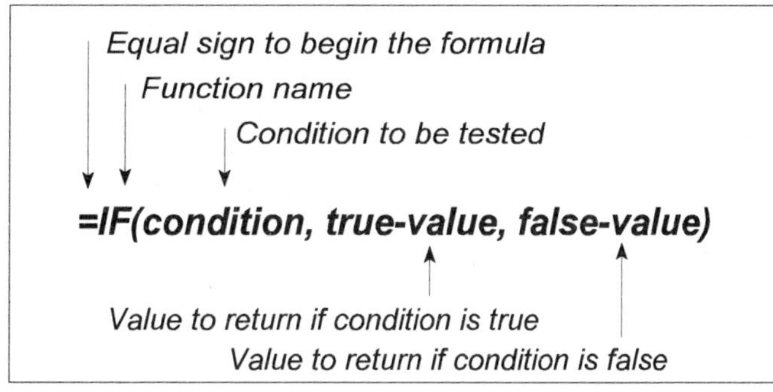

Figure 5.1

A *condition* has the form:

Expression-1 Comparison Operator Expression-2

Expression-1 and Expression-2 are any valid Excel expressions composed of cell references, constants and functions. Essentially, an expression is a formula without the equal sign. Thus to test if cell A3 has a value of 5 the condition is: A3 = 5. Here the first expression is a simple cell reference, while the second one is a constant. An example of a more complicated condition would be: (A1+A2)*10 > B1/B2.

Note: These *comparison operators* are generally called *relational operators* in computer science.

The comparison operators are:

=	equal to
>	greater than
>=	greater than or equal to
<	less than
<=	less than or equal to
<>	not equal to.

Some examples of IF formulas:

(a) =IF(A2<0, "Negative", "Positive")
Returns the text 'Negative' if A2 has a value less than 0, otherwise it returns 'Positive'.

(b) =IF(A10−B10 <= 0.001,0,1)
Returns 0 if (A10−B10) is less than or equal to 0.001, otherwise it returns 1.

(c) =IF(ABS(A10−B10)<=EPSILON, A10, B10)
The value in A10 is returned when the absolute value of (A10−B10) is less than or equal to the value stored in a cell called EPSILON, otherwise the value in B10 is returned.

(d) =IF(SUM(A12:A20)>0, SUM(A12:A20), "Error")
If the sum of the range is greater than 0, that value is returned, otherwise the text 'Error' is displayed.

(e) =IF(D2<0, NA(), D2)
When D2 is negative, the function NA() causes the Excel value #N/A (meaning 'not available' or 'not applicable') to be displayed. Sometimes we use this to mean 'Display something is wrong'.

(f) =IF(A1, "True", "False")
This will return the textual value 'True' if A1 contains a non-zero value, a formula giving a non-zero value, or the TRUE value. If A1 is empty, has the value 0, or the FALSE value then the text 'False' is returned.

IF functions may be *nested*. This means that within one IF function, we may use another IF function for either or both returned values.

(a) =IF(A1>10, IF(A1>50, "Big", "Medium"), "Small")
It is clear that if the condition A1 > 10 is false then the first IF returns 'Small'. What happens if the condition is true? The second IF comes into play. When A1 >100, the inner IF returns 'Big', otherwise it returns 'Medium'.

(b) =IF(A1>10, IF(A1>50, "Big", "Medium"), IF(A1<0, "Negative", "Small"))
Here both the true-value and the false-value of the outer IF are themselves IF functions.

Nesting up to seven levels is permitted provided the total number of characters in the cell does not exceed 1024. Remember you may use spaces in the formula to make it more readable.

No formula may contain more than 1024 characters. Facts like this may be confirmed by searching for *specifications* in Help.

A formula may be constructed using just a condition. Such formulas will return the Boolean values TRUE and FALSE. Some simple examples are shown in Figure 5.2 in which the left-hand side displays the formulas while the right-hand side displays the values. Row 3 demonstrates that TRUE and FALSE values are numerically equivalent to 1 and 0, respectively.

Excel 2002: When a cell contains the value TRUE, Excel 2002 may add a *Smart tag.* The lower right corner of the cell displays a purple triangle and, when the cell is active, a smart tag icon appears. Excel is set to tag financial symbols and TRUE is the NASDAQ symbol for the high-tech company TimeTrue, Inc. These tags may be disabled using Tools|AutoCorrect Options and opening the Smart Tag tab. Remove the check mark from *Smart tag list (MSN Money Financial Symbols).* See Help for more details.

	A	B	C
1	2	=4/2	4
2	=A1=B1	=C1>B1	=C1<=B1
3	=5*A2	=5*B2	=5*C2

	A	B	C
1	2	2	4
2	TRUE	TRUE	FALSE
3	5	5	0

Figure 5.2

The *logical functions* AND(), OR() and NOT() have the syntax AND(***logical1***, *logical2*,...) and OR(***logical1***, *logical2*,...), respectively. The arguments *logical1*, *logical2*, etc. are conditions that can have true or false values. The ellipses indicate that there can be up to 30 arguments. The AND function returns the TRUE value when all the arguments are true. The OR function returns TRUE when one or more of the arguments is true. These functions may be used in their own right or to construct compound conditions. Some examples follow. In Exercise 5 we look at these functions further.

(a) =IF(AND(A2>0, A2<11), A2, NA())
The value A2 is returned if A2 is greater than 0 and less than 11. Otherwise, the function NA() returns the error value #N/A.

(b) =IF(OR(A2>0, B2>A2/2), 3 , 6)
Returns the value of 3 if either A2 > 0 or B2 > A2/2. If neither condition is true, the value 6 is returned.

(c) =IF(NOT(A2=0), TRUE, FALSE)
This is the same as IF(A2=0, FALSE, TRUE).

(d) =IF(NOT(OR(A1=1, A2=1)), 1, 0)
This is a somewhat contrived example. It returns 1 if both A1 and A2 have a value that is not 1.

Exercise 1: A What-if Analysis

Acme Inc. makes widgets which are tested before being sold. The testing gives two values, *P* and *Q*. The requirements are that *P* be at least 1.25 and *Q* be no more than 0.5. Using some sample data, Acme wishes to know how many widgets pass the tests and how the results would change if the specifications were to be altered slightly. For this exercise, we will use only 10 data sets but in a real case there might be hundreds. The results of this small sample of widgets are shown in Figure 5.3.

	A	B	C	D	E
1	Quality control				
2		pmin	1.25		
3		qmax	0.5		
4					
5	P	Q	P result	Q test	Two test
6	1.24	1.08	0	0	0
7	1.36	0.50	1	1	1
8	1.44	0.40	1	1	1
9	1.57	0.54	1	0	0
10	1.09	0.82	0	0	0
11	1.52	0.65	1	0	0
12	1.23	0.75	0	0	0
13	1.65	0.62	1	0	0
14	1.24	0.36	0	1	0
15	1.05	0.55	0	0	0
16		Pass	50.0%	30.0%	20.0%

Figure 5.3

(a) Start a new workbook. On Sheet1 enter the text and values in A1:E5. Name the cells C2 and C3 as *pmin* and *qmax*, respectively. Enter the numeric values in A6:B15.

(b) Enter these formulas:
C6: =IF(A6>=pmin, 1, 0)
D6: =IF(B6<=qmax, 1, 0)
E6: =IF(AND(A6>=pmin,B6<=qmax), 1, 0)

Each of these formulas returns a value of 1 if a condition is met, otherwise they return 0. The first tests the *P* value, the second tests the *Q* value, and the third tests both values.

(c) Copy the three formulas down to row 15.

(d) Enter the text in B16 and right align it.

(e) In C16 enter the formula =SUM(C6:C15)/COUNT(C6:C15) and copy it to the two cells to the right.

(f) We are now ready to play the what-if game. By changing the value in C2, we can get answers to questions such as 'What percentage of widgets would pass if the acceptable value of *P* was (i) raised to 1.3 or (ii) lowered to 0.22?'

(g) Save the workbook as CHAP5.XLS.

Exercise 2: Avoiding Division by Zero

In Exercise 7 of Chapter 2 we developed a worksheet to find the effective value of four resistors in parallel. It was noted that we could not use the worksheet for less than four resistors by using a zero value since we would run into the divide by zero problem. On Sheet2 of CHAP5.XLS we will develop a worksheet to compute the effective value of up to six resistors in parallel using the formula $1/R = \sum 1/R_i$. Our final product will resemble Figure 5.4.

	A	B	C	D	E	F	G
1	Resistors in Parallel						
2							
3	Resistors	10	20	35	40	50	50
4	1/R	0.1	0.05	0.028571	0.025	0.02	0.02
5	R_e	4.105572	ohms				

Figure 5.4

(a) Enter the text shown in column A. In A5 type Re and in the formula bar select the letter *e* before using the command Format|Cells and selecting *Subscript* on the Font tab. Enter the numeric values shown in B3:G3.

(b) The formula in B4 is =IF(B3>0, 1/B3, " ") which returns the reciprocal of the resistance if its value is greater than 0. Otherwise, the formula returns a space. Copy this across to G4.

(c) In B5 enter =1/SUM(B4:G4). Check the result using pencil and paper. (OK, you can use your calculator!)

(d) Now let us see if it works for four resistors. Enter 25 for the value of the first four resistors and either leave the last two empty or use the value 0. Do you get 6.25 ohms? Save your workbook.

It was noted in Chapter 4 that the SUM function can tolerate non-numeric values in cells forming part of an argument. We would be in trouble if we did not know the SUM function and, in its place, used =B4 + C4 + D4 + E4 + F4 + G4. The simple formula will not tolerate non-numeric values and will return the #VALUE! error.

Exercise 3: Quadratic Equation Solver

Note: Problem 6 at the end of the chapter suggests another method for finding the smaller root.

In this exercise we design a worksheet to solve a quadratic equation in the form $ax^2 + bx + c = 0$ using the quadratic formula:

$$x = \frac{-b \pm \sqrt{b^2 - 4ac}}{2a}$$

The quantity $\sqrt{b^2 - 4ac}$ is called the discriminant because its value determines the number (0, 1 or 2) of real roots of the equation.

When this exercise is completed, the worksheet will resemble that in Figure 5.5.

	A	B	C	D	E
1	Quadratic Equation Solver				
2	a	b	c		disc
3	1	5	6		1
4					
5	Number of real roots		2		
6	Root 1	-2	Root 2	-3	

Figure 5.5

(a) Open CHAP5.XLS. On Sheet3 enter the values shown in A1:C3. Select A2:C3 and centre the entries with the button on the Formatting toolbar.

(b) With A2:C3 still selected use the command Insert|Name|Create to give the cells A3:C3 the names in the cells above them.

(c) Type disc (short for discriminant) in E2. In E3 enter the formula =b*b - 4*a*c_ in E3. Centre E3:E4 and create the name *disc* for the cell E3.

(d) Temporarily ignore the entries in A5, B5, A6 and C6 of Figure 5.5. Type these formulas in B6 and D6:
B6: =(-b + SQRT(disc))/(2*a)
D6: =(-b - SQRT(disc))/(2*a)

(e) Save the workbook CHAP5.XLS.

You now have an operational worksheet. Test it with quadratic equations whose roots you know. What happens if the value of the discriminant is negative? Cells B6 and D6 show the error value #NUM! since it is impossible to evaluate the square root of a negative number without entering the realm of imaginary numbers.

The next steps will improve the behaviour of the worksheet when the discriminant is negative and add some additional information. The results we are aiming for are shown in Figure 5.6.

Disc > 0

	A	B	C	D
5	Number of real roots		2	
6	Root 1	-2	Root 2	-3

Disc = 0

	A	B	C	D
5	Number of real roots		1	
6	Double Root	2		

Disc < 0

	A	B	C	D
5	Number of real roots		0	
6				

Figure 5.6

(f) Enter the text in A5. In C5 enter the formula =IF(disc<0, 0, IF(disc=0, 1, 2)). This returns 0 when the discriminant is negative, 1 when it is zero and 2 in all other cases.

(g) In A6 enter the formula =IF(C5>0,IF(C5=1, “Double Root”, “Root 1”), “ ”). If there is one root, this returns the text ‘Double Root’, if there are two identical roots it returns ‘Root 1’. When there are no real roots, it returns an empty test string.

(h) We require the formula in B6 to return a root when the discriminant has a zero or positive value, and an empty text string otherwise. We can achieve this by modifying it to read: =IF(disc>=0,(-b+SQRT(disc))/(2*a), " ").

(i) In C6 enter the formula =IF(C5=2, "Root 2", " ") to return the text 'Root 2' only when the discriminant has a positive non-zero value.

(j) Modify D6 to =IF(disc>0,(-b-SQRT(disc))/(2*a), " ") to return the value of the second root only when the discriminant has a positive non-zero value.

(k) Save the workbook.

Exercise 4: Protecting the Worksheet

It has taken a lot of work to get the worksheet in the previous exercise operating and we would like to protect it in such a way that a user can change the values in A3:C3 but nowhere else.

(a) Select the cells A3:C3 and use the command Format|Cells. In the Protection tab sheet click the box next to 'Locked' to remove the check mark.

(b) Use the command Tools|Protection|Protect Worksheet to protect the worksheet. Do not use a password unless you are totally confident that you will remember it – and then write it down and hide the paper in your sock drawer because, while password-breaking programs are readily available using an Internet search, they are expensive. A user may now change only the values in cells A3:C3.

(c) Perhaps you would rather the user did not see the entries in E2:E3. To make any changes you must first unprotect the worksheet.

(d) You could either hide column E or just hide the entries in E2:E3. See if you can find out how to do either. Reprotect the worksheet when you are finished.

If you wish to give this worksheet to someone else here is a simple way. Open CHAP5.XLS and select every cell by clicking on the top left of the worksheet where the row and column headings meet. Open a new workbook and with A1 of Sheet1 the active cell, click on the Paste tool. You now have a duplicate ready to gift wrap.

Exercise 5: Imaginary Roots

If you are familiar with imaginary (or complex) numbers, you may wish to modify the worksheet of Exercise 2 to compute the imaginary roots when the discriminant has a negative value. The two roots, which are complex conjugates, are given by $-b/2a \pm i\sqrt{-(b^2-4ac)}/2a$.

(a) We could modify the work on Sheet3 of this workbook but you may wish to keep that sheet unchanged. To save retyping everything we will make a duplicate of that sheet. Hold down Ctrl and drag the Sheet3 tab to the right. You will see an icon representing a sheet bearing a + symbol. When you release the mouse button there will be a new worksheet called Sheet3(2).

(b) On the duplicate sheet, enter the following formulas which will display text only when the discriminant has a negative value.
A8: =IF(disc<0, "Imaginary roots", " ")
A9: =IF(disc<0, "Root 1", " ")
A10: =IF(disc<0, "Root 2", " ").

(c) In B9 we shall use the COMPLEX function. If A1 has the value 2 and A2 has the value 3, then the formula =COMPLEX(A1,A2) would return 2+3i. So for the first root we could use: COMPLEX(-b/(2*a), SQRT(-disc)/(2*a)). However, this will give an error value unless the discriminant has a negative value, so we modify it to =IF(disc<0, COMPLEX(-b/(2*a), SQRT(-disc)/(2*a)), " ").

This would be fine for most cases but sometimes Microsoft Excel will evaluate the terms to so many decimal places that the result will be unreadable. Using the ROUND function will solve this problem. In B9 carefully type the formula:
=IF(disc<0,COMPLEX(ROUND(-b/(2*a),2), ROUND(SQRT(-disc)/(2*a),2)), " ").

(d) We could use a similar formula in B10, changing the imaginary part to -SQRT(disc)/(2*a). But since the second root is the complex conjugate of the first we will use: =IF(disc<0, IMCONJUGATE(B9), " ").

(e) Save the workbook.

Test your worksheet with $a = 1$, $b = 0$ and $c = 9$. The imaginary roots are 3i and −3i. Do a paper and pencil calculation to find the roots of $x^2 - 2x + 10 = 0$ and test your worksheet.

Exercise 6: Logical Functions

At the start of this chapter, we saw some examples of IF formulas using NOT, AND and OR. These terms are the logical, or Boolean, functions. They return the values TRUE or FALSE.

(a) Insert a new sheet, Sheet4, in CHAP5.XLS. Enter the values and formulas as shown in Figure 5.7 and note the values returned.

	A	B	C
1	10	=A1=10	=AND(A1>5,A2>5)
2	20	=A3>20	=OR(A1>15,A2>15)
3	30	=NOT(A1=10)	=A3=A1+A2

Figure 5.7

(b) The logical functions may be combined to make more complex tests. You should be very careful when doing this because (i) you must understand the following order of precedence: NOT followed by AND then OR; (ii) in everyday language we frequently say 'and' where 'or' is what is really meant. Logically, it is meaningless to say 'This is a list of my friends who live in Canada and the USA' because it is unlikely that any friend has a residence in both countries. If in doubt use two or more cells to make complex logic tests.

Note: To enter the value TRUE (or FALSE) you may type:

(i) TRUE
(ii) =TRUE, or
(iii) =TRUE().

Use the last form when you want your Excel worksheet to be compatible with other spreadsheets.

There are some common combinations that are useful to know. In the following table, *A* and *B* may be expressions or references to cells containing the values TRUE or FALSE.

Name	Formula	TRUE returned if
NAND	=NOT(AND(A,B))	*Not both* true
NOR	=NOT(OR(A,B))	*Neither* is true
XOR	=OR(AND(A, NOT(B)),AND(B, NOT(A)))	*Only one* is true

Construct a worksheet to test the formulas shown. Show that the statement 'Not both true' is *not* equivalent to 'Both are false.'

Table Lookup Functions

Table lookup functions have a range of uses. Whenever you find yourself composing a multi-nested IF function you should consider whether a lookup function would be more appropriate. A vertical table has its headings in a row while a horizontal one has them in a column. There are no inherent advantages of one over the other.

The functions VLOOKUP and HLOOKUP have similar syntax:

VLOOKUP(***lookup_value, table_array, column_index_num,*** *range_lookup*)

HLOOKUP(***lookup_value, table_array, row_index_num,*** *range_lookup*)

Lookup_value	Is the value to be located in the first column of a vertical table (or the first row of a horizontal table). *Lookup_value* may be either a numeric or text value or a cell reference.
Table_array	Is the range reference or name of the table.
Column_index_num (*row_index_num*)	Is the column (or row) of the table from which the value is to be returned.
Range_lookup	Is a logical value (TRUE or FALSE) specifying whether you want an approximate or an exact match. If range_lookup is TRUE or omitted, and there is no exact match, then the function returns the next largest value that is less than the lookup value. If FALSE and no exact match is found, the function will return the error value #N/A. If lookup_value is less than the lowest value in the first column (first row with HLOOKUP), the function returns the #N/A error value.

The MATCH function returns the relative position of a lookup_value in an array. Its syntax is: MATCH(***lookup_value, lookup_array,*** *match_type*). The first two arguments have the same meaning as above. Use 1 for *match_type* when the table is sorted in ascending order and you wish to find the largest value that is less than or equal to *lookup-value*. Use 0 when you needed an exact match; the table need not be sorted. Use −1 when the table is sorted in descending order and you wish to find the smallest value that is greater than or equal to *lookup-value*. When *lookup-value* is non-numeric, MATCH, VLOOKUP and HLOOKUP are not case sensitive. MATCH may also be used with wildcards.

The INDEX function returns an element from an array and has two forms. The syntax of the first form is INDEX(***array,*** *row_num, column_num*). Thus =INDEX(A1:C10, 2, 3) returns the value at the intersection of row 2 and column 3 of the table A1:C10. In this example, it returns the value from cell C2.

Exercise 7: Horizontal Lookup

For the purpose of this exercise, a geologist wishes to grade some ore samples based on their rare metal content. Ore with 50 to 59 ppm is to be given a low grade, from 60 to 79 merits a medium ranking, from 80 to 99 is considered high and anything above that is very high. Our complete worksheet will resemble Figure 5.8.

	A	B	C	D	E	F	G	H
1	Horizontal table lookup example					Lookup Table		
2				ppm	50	60	80	100
3				Grade	low	medium	high	very high
4	Site	ppm metal	Grade					
5	A	75	medium					
6	A	56	low					
7	B	86	high					
8	B	60	medium					
9	C	34	#N/A					
10	C	120	very high					

Figure 5.8

(a) On Sheet5 of CHAP5.XLS, enter the text and values shown in A1:H3, A4:C4 and A5:B10.

(b) In C5, enter =HLOOKUP(B5, $D$2:$H$3, 2, TRUE) and copy this down to C10.

The formula in C5 attempts to locate the value in B5 in the first row of the range D2:H3. If an exact match is found, it returns the value from the corresponding cell in row 2 (the third argument) of the table. Otherwise it locates in the first row the largest value that is not greater than B5 and returns the corresponding cell from row 2. The results are correct other than the #N/A value in C9. Perhaps we would rather have an empty cell when the *ppm* value is less than 50.

(c) To accomplish this, we change the formula in C5 to read =IF(B5>=50, HLOOKUP(B5, $D$2:$H$3, 2, TRUE), " ") and copy this down to C10.

It might have been better to give the table a name, say OreTable, and then we would use =IF(B5>=50, HLOOKUP(B5, OreTable, 2, TRUE), " ") in C5. An alternative to the IF construct would be to insert a new column in the table with 0 in row 1 and an empty cell in row 2.

The author's web site contains a workbook with a larger vertical table. This can be used to compute the formula mass and percent composition of any chemical when its empirical formula is entered.

Exercise 8: Vertical Lookup

In this example, a nutritionist enters a client's height, frame type and weight, and the worksheet gives the person's optimal weight and a comment on his actual weight. To keep the exercise to a reasonable size, we limit ourselves to male clients. Our final product will resemble Figure 5.9.

	A	B	C	D	E	F	G
1	Optimal male weight			height	S	M	L
2				157	59	61	65
3	Height	159		160	60	62	66
4	Frame	L		163	61	63	67
5	Weight	70		165	62	64	69
6				168	63	65	70
7	Optimal	65 kg		170	64	67	72
8	Comment	5 kg	over	173	65	68	73
9				175	66	70	75
10				178	67	71	77
11	Match functions			180	68	72	78
12	frame	3		183	70	74	79
13	height	1		185	71	76	81
14				188	73	77	83
15				191	75	79	86
16				193	76	81	88

Figure 5.9

(a) On Sheet6 of CHAP5.XLS begin by entering the text shown in A1:A13. Then enter the table in D1:G16.

(b) Use Insert|Name|Define to name the following areas of the table: D2:D16 to be named *height*; E1:G1 to be named *frame*, and E2:G16 to be named *weight.*

(c) Enter the values shown in B3:B5. These are the client's data.

(d) Jump down to B12 and enter =MATCH(B4,frame,0). This will search for the client's frame type (the value in B4) in the array called frame (E1:G1) and return the matching position. When the client's frame type is L (large) the formula should return the value 3.

(e) In B13 enter =MATCH(B3,height,1). This will search for the client's height in the height table (D2:D16) and return the position of closest match. When the client's height is 162.5 cm the returned value should be 3.

(f) Now we know the row and the column to use to locate the optimal weight for a male having this height and frame type using the INDEX function. The formula in B7 is =INDEX(weight,B13,B12). If you have followed the exercise so far then you should be able to see that we do not really need the formula in B12 and B13. The formula =INDEX(weight, MATCH(B3,height,1), MATCH(B4,frame,0)) would work in B7.

(g) Next we will format B7 to show a value and its units. With B7 as the active cell use Format|Cell and select Custom from the Category area. In the text box enter 00 "kg" – see Figure 5.10. The formula in B8 is =IF(B5=B7, "OK", ABS(B7-B5)&"kg"). This shows another way of adding units but this time we need to change the justification to left since the cell has a textual value and will, by default, be left aligned.

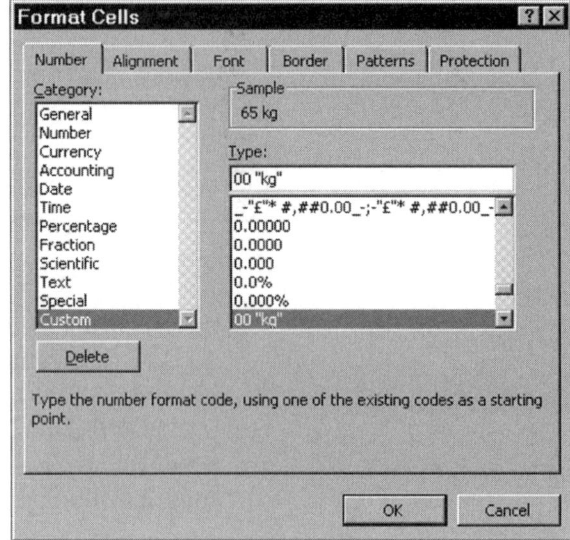

Figure 5.10

(h) Complete the worksheet with =IF(B5=B7, " ",IF(B7<B5, "over", "under")) in C8.

(i) Save the workbook.

Experiment with other values for the client's height, frame type and weight. Have you found a fault? With a height of 159, the worksheet uses the first row of the table. Should it not use the second since 159 is closer to 160 than it is to 157? This is left as an exercise for the reader to solve.

Exercise 9: Conditional Summing and Counting

The SUMIF and COUNTIF functions are useful for analysing lists of data. The syntax for SUMIF is: SUMIF(***range***, ***criteria***, *sum_range*). To sum the values in the range A1:A100 that are greater than 5 we would use =SUMIF(A1:A100, “>5"). The array to be summed may differ from the array used for the selection. To sum all the values in column B when the corresponding cells in column A have the value ‘OK’ we would use =SUMIF(A1:A100, “OK”, B1:B100). The COUNTIF function has the syntax COUNTIF(***range***, ***criteria***).

We will do some simple experiments with these functions. Our complete worksheet will resemble Figure 5.11.

	A	B	C	D	E	F	G
1	SUMIF & COUNTIF						
2							
3	8		Sum if > 10	210		Countif > 10	12
4	19						
5	13		Sum if less than			Count if less than	
6	17		5	8		5	3
7	9		10	48		10	8
8	3		15	83		15	11
9	19		20	228		20	19
10	19		25	268		25	21
11	10						

Figure 5.11

(a) We need a list of numbers to practise with. Rather than type them we will use the random number generator. On Sheet7 in A3 enter =RANDBETWEEN(1,20). If you get a #NAME? error, you need to install the Analysis ToolPak. Copy this formula down to A23.

(b) The random numbers will change every time we change the worksheet. To prevent this we will turn the formulas into values. Select A3:A23, click the Copy tool, use the command Edit|Paste Special and click the *Values* option button. Click OK on the Paste Special dialog box and on Esc to close copy mode. Check that A3:A23 contains values not formulas.

(c) The formula in D3 is =SUMIF(A3:A23, “>10”). The formula in G3 is =COUNTIF(A3:A23, “>10”).

(d) In D6 enter =SUMIF($A$3:$A$23, “<”&C6). The absolute references are so that you can copy this down to D10. In G6 enter =COUNTIF($A$3:$A$23, “<”&F6) and copy this down to G10.

Exercise 10: Array Formulas

We have seen some functions which use an array for their input. The simplest of these are SUM and COUNT. In Exercise 8 of Chapter 4 we used the MMULTI function which uses an array for input but also produces an array of values as output. It is also possible to create from simple functions formulas that input arrays. This is a very broad topic so we can only touch upon it in the next example.

You may wish to review Exercise 8 of Chapter 4 before continuing. For example, have you remembered that we must use Ctrl+⇧Shift+Enter↵ to complete an array formula entry?

The Boolean functions AND and OR do not act in the expected way when used in array formulas- they always return a single value, never an array.

For more array-formula examples including interesting uses of OR visit www.emailoffice.com/excel/arrays-bobumlas.html

To investigate array formula we will imagine we have some data and we need the following quantities: (i) sum of the cubes, (ii) a count of how many values lie in a ceratin range (let this be 5 to 10), (iii) the sum of the squares of values that lie within that range, and (iv) the sum of the *N* largest values. Generally one would use an array formula only with a large data set but it is more convenient to demonstrate it on a small set to show how one can check that the array formula has been constructed correctly.

(a) Set up a worksheet as shown in Figure 5.12. Enter the values in A2:A10 and name this range as *data*.

	A	B	C	D
1	data	cube	range count	range sum
2	2	8	0	0
3	3	27	0	0
4	5	125	1	25
5	6	216	1	36
6	10	1000	1	100
7	11	1331	0	0
8	9	729	1	81
9	6	216	1	36
10	5	125	1	25
11	SUM	3777	6	303
12				
13	Sum of cubes		3777	
14	Count of values in range		6	
15	Sum of values in range		303	
16	Sum of 3 largest		30	

Figure 5.12

(b) In row 2 enter these formulas and copy them down to row 10:
B2: =A2^3
C2: =IF(AND(A2>=5), A2<=10, 1,0)
D2: =IF(AND(A2>=5), A2<=10, A2^2,0)

In row 11, use the SUM function to sum the B, C and D columns.

(c) Now we see that we can obtain these summations without the intermediate data. The formulas in the lower part of the sheet are:

C13: {=SUM(data^3)}
C14: {=SUM((data>=5)*(data<=10))}
C15: {=SUM((data>=5)*(data<=10)*data^2)}
C16: {=SUM(LARGE(data,{1,2,3}))}

Remember to enter the formulas without the braces and complete each one with Ctrl+Shift+Enter.

Observe how the results from the first three agree with the summation in line 11. We could have found the sum of the three largest values with the non-array formula =LARGE(data,1) + LARGE(data,2) + LARGE(data,3). However, had we wanted the sum of the 10 largest, an array formula would have been more convenient.

It is instructive to investigate how Excel performs these calculations. Read all of the next step before you proceed.

(d) Select C13 and in the formula bar use the mouse to select the *data^3* portion of the formula. Press F9 to calculate this part of the formula. Excel responds by displaying =SUM({8;27;125;216;1000;1331;729;216;125}). Excel has generated an array in which each item is the cube of the corresponding item in the *data* range. Press Esc to undo the evaluation; use the Undo tool if you forget to do this. Experiment with portions of the other formulas.

Users of Excel XP have access to a tool that performs these evaluations more conveniently. Make C15 the active cell and use the command Tools|Formula Auditing|Evaluate Formula. Click the *Evaluate* button on the dialog box and observe the result in the window. After four evaluations we will have =SUM({FALSE;FALSE;TRUE;TRUE;TRUE;TRUE;TRUE;TRUE;TRUE}*{TRUE;TRUE;TRUE;TRUE;TRUE;FALSE;TRUE;TRUE;TRUE}*data^2). The seventh evaluation gives =SUM({0;0;1;1;1;0;1;1;1}*{49;9;25;36;100;121;81;36;25}). The next evaluation gives =SUM({0;0;25;36;100;0;81;36;25}) – note how this matched the data in D2:D10. Finally we get the result 303.

Problems

1.* To compute the value of sin(*x*)/*x*, a student uses =SIN(A1)/A1. However, this will not give the correct value of 1 when A1 has a value of zero. Correct the formula.

2.* In the worksheet shown below, cells A2:C2 have the names displayed above them. A formula is required in D2 to display 'Pass' or 'Fail' depending on the values of *x*, *y* and *z*. A 'Pass' value requires (i) that *x* be less than 2, (ii) that *y* has a value between 12 and 14 inclusive, and (iii) *z* is not negative. Complete the following using IF three times to achieve this result: =IF(x < 2, IF(y ...

	A	B	C	D
1	x	y	z	Result
2	1.9	12	0	Pass

3.* Simplify the formula in 1 above using AND twice.

4.* This problem is for electrical engineers. (a) Using IF functions without AND or OR, show that the output (F) in the *voter circuit* below is YES (1) only when the majority of the inputs are YES. (b) Can you suggest a method using AND and OR without IF?

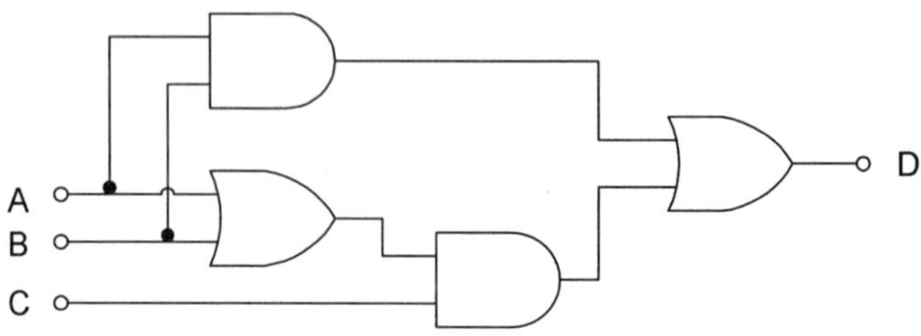

	A	B	C	D	E	F	G
1	A	B	C	First AND	First OR	Second AND	Second OR
2	0	0	0	0	0	0	0
3	0	0	1	0	0	0	0
4					etc...		

5.* The worksheet depicted in the figure below can be used to compute the value of a resistor from its coloured bands. The user enters the abbreviation for the colours in row 19. Row 21 serves the purpose of showing the user how the abbreviations were interpreted. Row 22 returns the first and second digits and the multiplier. Row 3 displays the resistor's value and tolerance. What are the formulas in rows 21, 22 and 23?

	A	B	C	D	E	F
1	Resistor codes					
2			1st digit	2nd digit	multiplier	tolerance
3	bk	black	0	0	1	1%
4	br	brown	1	1	10	2%
5	r	red	2	2	100	3%
6	o	orange	3	3	1000	4%
7	y	yellow	4	4	10,000	error
8	g	green	5	5	100,000	error
9	b	blue	6	6	1,000,000	error
10	v	violet	7	7	10,000,000	error
11	gr	gray	8	8	100,000,000	error
12	w	white	9	9	1.00E+09	error
13	gold	gold			0	5%
14	silver	silver			0	10%
15	none	no colour				20%
16						
17						
18	Band 1	Band 2	Band 3	Band 4		
19	r	r	o	none		
20						
21	red	red	orange	no colour		
22	2	2	1000			
23	Value	22000	20%			

6. Textbooks on numerical computing quite correctly warn of the round-off errors that can occur when subtracting two numbers that are very close in size. When the traditional method is used to solve a quadratic equation where $b^2 >> 4ac$, the root of smallest magnitude is found by subtracting from b a number that differs little from it. To avoid this, we first find the quantity $Q = -(b+sign(b) \times sqrt(b^2 - 4ac))/2$, and then compute the roots as $x_1 = Q/a$ and $x_2 = c/Q$.

   Construct a worksheet similar to that shown below to compare the two methods. You will need to look in Help for information on the SIGN function. Use an IF statement in B8 and B9 such that B8 contains the numerically larger root to facilitate comparison with the alternative method.

	A	B	C	D	E	F	G	H
1	Quadratic Equation Solver							
2								
3	a	b	c					
4	1	256	1.25E-06					
5								
6	Traditional				Alternative			
7	disc	6.55360000E+04			Q	-2.56000000E+02		
8	Root 1	-2.56000000E+02	Test 1	-2.2422E-12	Root 1	-2.56000000E+02	Test 1	-2.2422E-12
9	Root 2	-4.88282126E-09	Test 2	-2.2421E-12	Root 2	-4.88281250E-09	Test 2	0.0000E+00

6 Charts

Concepts

This chapter shows how to create graphs from data in a worksheet. So why does the chapter have the title *Charts*? Whereas business people show charts at their meetings, scientists and engineers use graphs to display their data. The term you use depends on your occupation but the object is the same. Since the largest market for Microsoft Excel is the business world, it uses the term charts. We will do the same so that you will remember to look under this term in the Help facility.

Types of Charts

Microsoft Excel offers the user some 300 chart formats including Line, XY (Scatter), Column, Bar and Pie charts. Some of the chart types are shown in Figure 6.1. We concentrate on XY charts since these are generally the most useful for scientists and engineers. When you have mastered this chapter, you will be able to create charts of other types with a little experimentation.

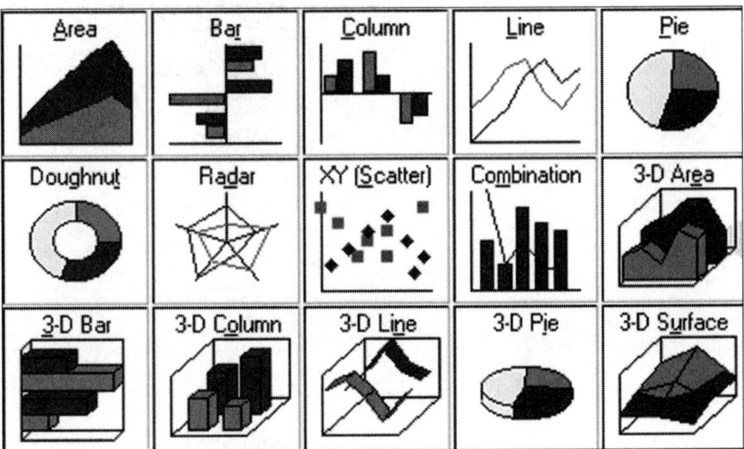

Figure 6.1

Line and XY (Scatter) Charts

Many Excel beginners are confused by the two terms *Line chart* and *XY (Scatter) chart.* Let us agree to drop the *scatter* part of the name; this is a term from statistics and generally has no meaning to the scientific or engineering users. The diagrams in Figure 6.1

add to the confusion about a *Line* and an *XY* chart. Both types are capable of displaying their data in line or marker format. Indeed, we shall see that you can use both lines and markers in a chart.

So how do the two types differ? The XY chart is another name for the *graph* (or more correctly the *coordinate graph*) with which we are familiar. This is what we need to plot ordered pairs of data to observe the dependence of one value on another.

Moral: When the x-values are numeric you normally require an XY chart.

A Line chart is generally used when the x-values are textual (days of the week, names of companies, places, etc.). Furthermore, even when the x-values in a Line chart are numbers, they are treated as text. Each x value is placed one after the other on the x-axis, regardless of value. It takes practice to make a Line chart with numeric x-values since Excel will think there are two y-value ranges. The trick is to make an XY chart first and convert it to a Line chart. By default, the data points are between the tick marks on the x-axis but this can be changed.

In Figure 6.2 we have charts produced from two data sets. The XY and Line charts made from the first set are very similar but observe the difference when the x-values have non-regular intervals.

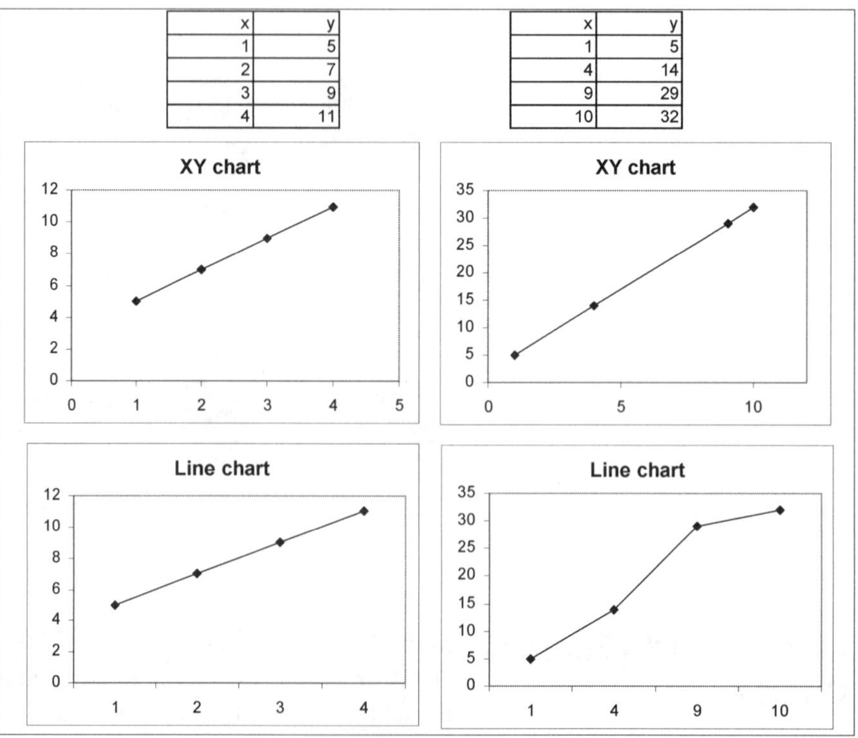

Figure 6.2

Embedded Charts and Chartsheets

A chart may be created either on a worksheet or on a separate chartsheet. In the former case we say the chart is *embedded.* We shall make all our charts embedded since it is then easier to see how the chart changes when the data is altered. The steps for creating and modifying a chart in its own sheet are essentially the same as for an embedded chart. This is explained in Exercise 1.

Anatomy of a Chart

Objects

You know that the *x*-axis is the horizontal axis while the *y*-axis is vertical. You are also familiar with the terms *titles*, *legends* and *gridlines*. However, the terms *chart area* and *plot area* may be new. Their meaning is demonstrated in Figure 6.3. Collectively, these items are called *objects*. The chart area is everything within the border. The plot area is everything other than the titles and legend. In this chart, the plot area has a pattern. If you do not need the pattern you must set pattern to None, as is explained in a later exercise. Gridlines are optional; you may include vertical and/or horizontal gridlines.

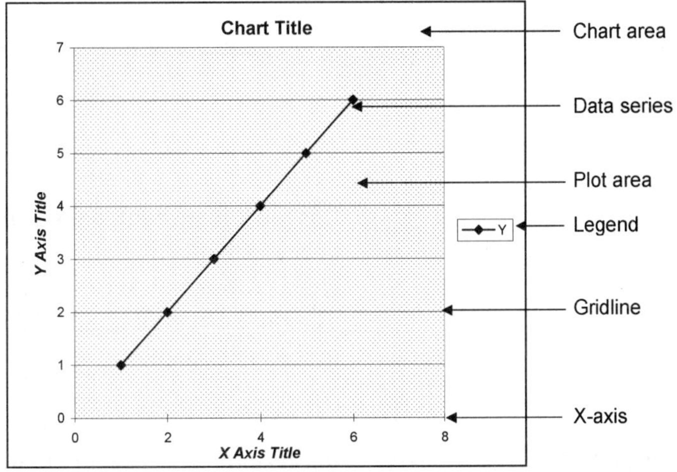

Figure 6.3

Markers and lines

The data in this chart is plotted with both *markers* and a *line*. You may opt to have one or both of these. You can also specify neither markers nor line in which case the data disappears! The colour of each may be specified separately. There are a number of styles (e.g. solid or dotted) and width options for the line. Similarly, the shape of the markers may be changed. The scale (maximum and minimum values) of each axis may be changed, and there are options to change the number and placement of the *tick marks* on each axis.

Smoothing Option

One of the options for changing the appearance of a chart is *smoothing*. This option may be selected as the chart is made or later using either the *Chart type* or the *Format Data Series* dialog. In most cases we will want to smooth our data but there may be occasions when it is inappropriate to smooth experimental data. Figure 6.4 shows the effect of this option when plotting the function $y = 3x^2 + 4$. You may wish to experiment with this feature when working through the exercises.

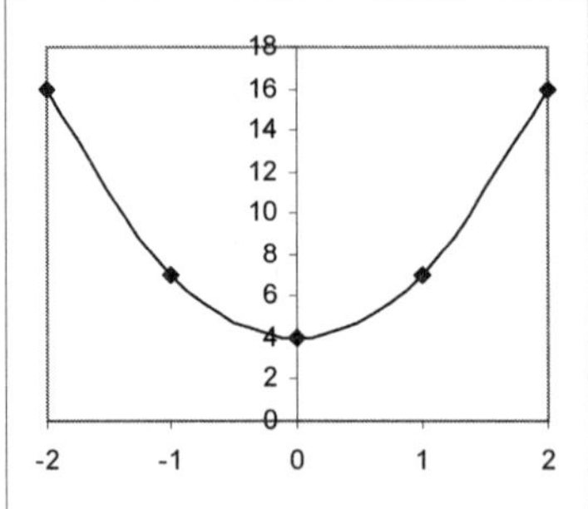

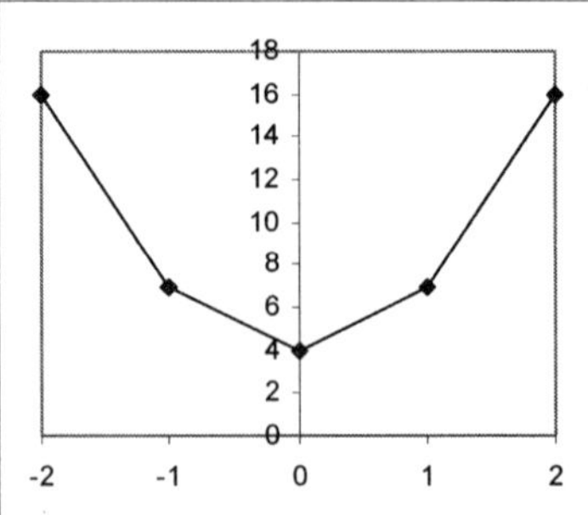

Figure 6.4

It is most important that you understand that the smooth line is not necessarily the line of best fit. In general the smooth line is a *cubic spline* which is a series of curves constructed to join seamlessly. The line of best fit is called the *trendline* in Excel. We explore this in the next chapter.

Exercise 1: Creating an XY Chart

In this exercise we will create an XY chart. Figure 6.5 shows the data we will plot and the chart that will result.

(a) Start a new workbook. In the range A1:B6 type the data shown in Figure 6.5.

(b) Select the range A1:B6. Click the Chart icon which is on the Standard toolbar. This will invoke the Chart Wizard – a series of four dialog boxes that help you construct the chart. You may later change any of the options selected in these four steps.

(c) In Step 1 of the Chart Wizard (Figure 6.6) we specify the type and subtype of chart. For this exercise we require an XY type chart. For the subtype, select chart with smoothed lines and markers. Press the *Press and Hold to View Sample* button to get a thumbnail sketch of how the chart will appear.

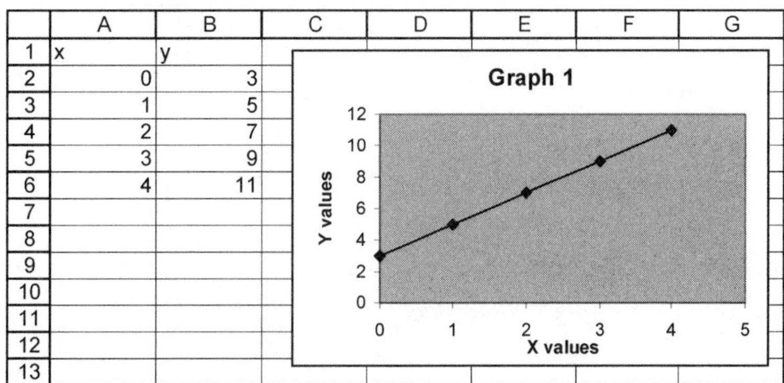

	A	B
1	x	y
2	0	3
3	1	5
4	2	7
5	3	9
6	4	11

Figure 6.5

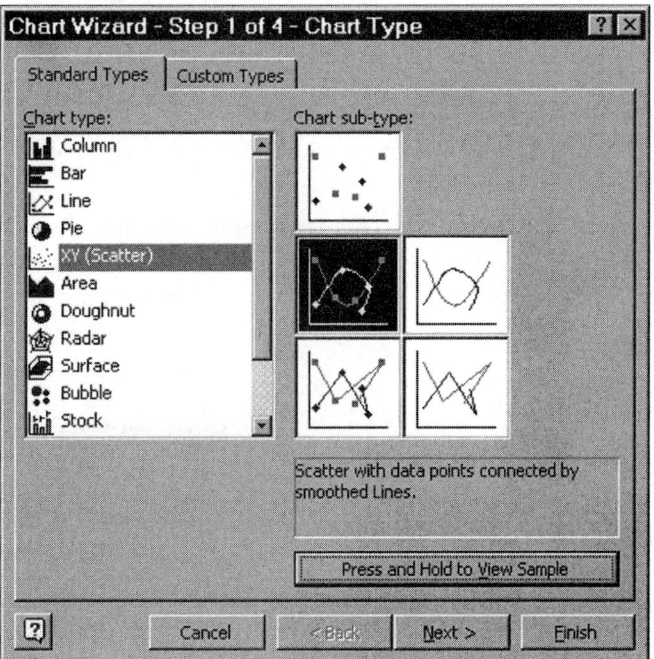

Figure 6.6

(d) Click the *Next* button to bring up Step 2 as shown in Figure 6.7. Generally, one need do nothing in Step 2 except press the next button.

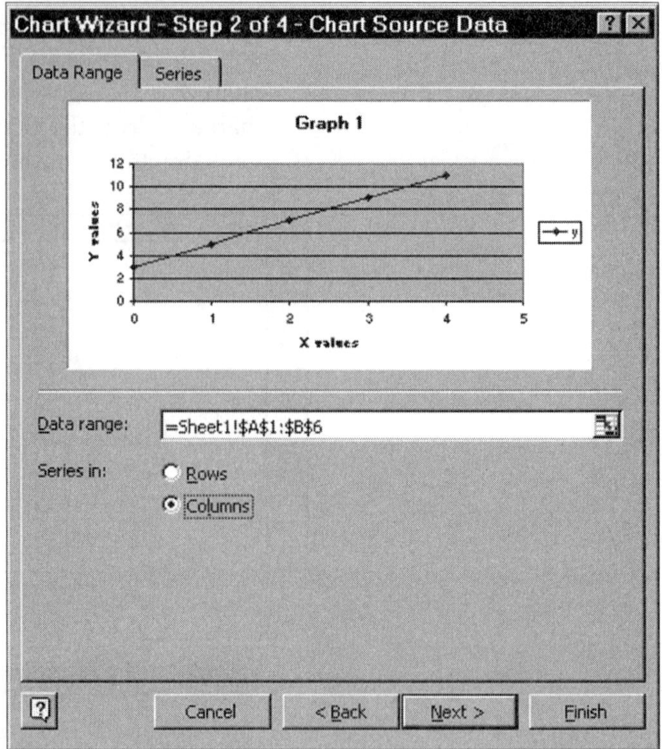

Figure 6.7

(e) Step 3 of the Chart Wizard (Figure 6.8) offers a number of options. We began the exercise by selecting a range of data together with headers; Excel has taken the text in the second column as the title. Click in the *Title* box and make the title *Graph 1.* Add the two axes titles.

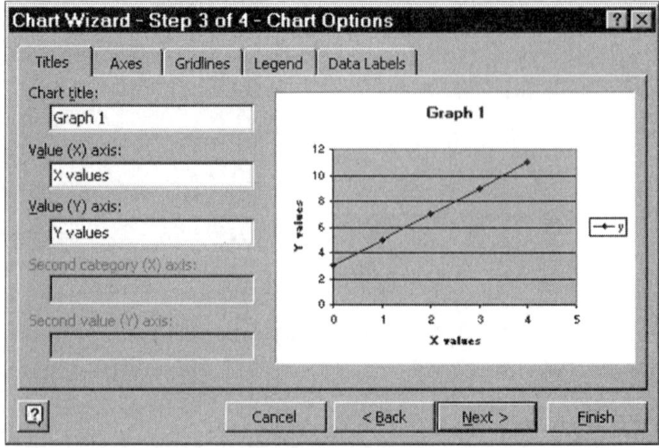

Figure 6.8

(f) Open the *Gridline* tab of the dialog box and click the appropriate boxes to remove all gridlines. If you have jumped ahead to Step 4, use the Back button to return to Step 3. Now open the *Legend* tab and click the box to remove the legend. A legend is redundant when there is only one data series.

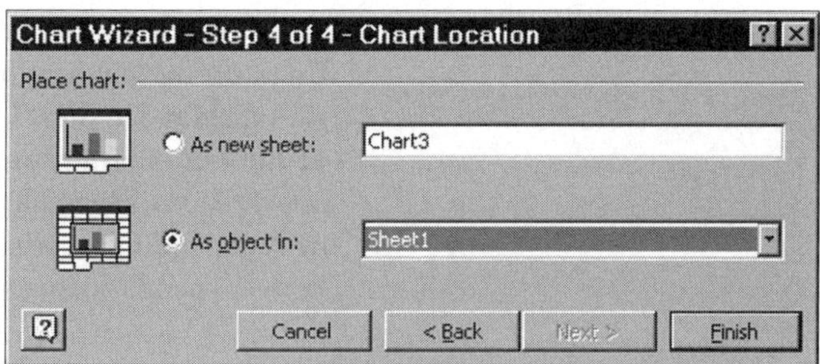

Figure 6.9

(g) Use the *Next* button to proceed to Step 4 (Figure 6.9). It is here that we may specify if the chart is to be embedded in the worksheet or to be placed on a separate chartsheet. We will opt for the former. Press *Finish* to complete the Chart Wizard.

At this point the chart is displayed with eight solid boxes (*handles*) around its border. When they are showing, we speak of the chart as being *selected* or *activated.*

A Chart toolbar is probably visible. If not, use View|Toolbars and click on Chart so that this toolbar appears whenever a chart is selected. We will use this in the next exercise.

Tips: If you hold down Shift when dragging a chart, it can only be moved parallel to either the column or the row headers. Holding down Alt ensures the chart is aligned with a cell border when it is moved or resized. These 'tricks' help when you want to align several charts or make them of equal size.

A quick way to copy a chart is to hold Ctrl while dragging a chart.

Range finders: When a chart is selected, the data used to make the chart is outlined in colour-coded rangefinders – see Figure 6.10.

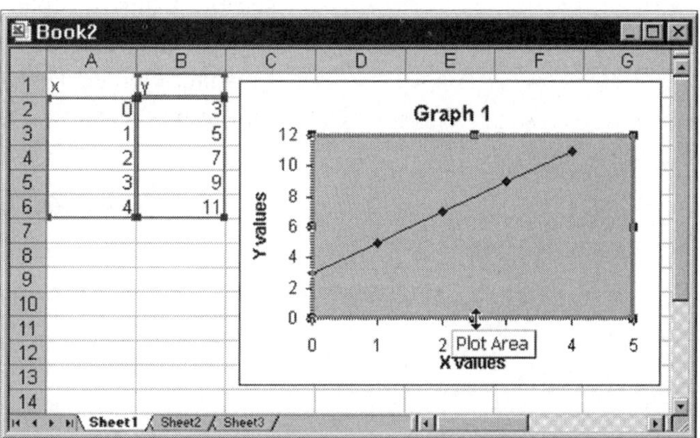

Figure 6.10

(h) In all likelihood, you will be disappointed with the product. Do not despair, it can quickly be made very acceptable. If the handles are absent, click just inside the chart.

Hold down the mouse button and drag the chart to the desired position. Pull any of the handles to resize the chart.

Click on the inner border (the border of the plot area) and it will display handles. Carefully drag the handles in the centre of the bottom and top borders (Figure 6.10) to improve the appearance of the chart. The maximum *x*-axis value will be 5 or 6, depending on what size you make your chart. We will see how to specify a constant value in the next exercise.

Exercise 2: Modifying a Chart

We saw at the end of Exercise 1 how to change the position of a chart, and size of the chart and the plot areas. In this exercise we will make other changes to the chart created in that exercise. The process of altering the appearance of objects on a chart is called formatting.

1. Format Plot Area

By default, Microsoft Excel gives the plot area a grey shaded background. While this looks fine on the screen, generally it is less pleasing on a printed page. We will remove this so that the plot area has the same appearance as the chart area.

(a) Click anywhere within the plot area. If you are not sure that you have the correct area, let the mouse rest for a few seconds and a screen tip displaying *Plot Area* will appear. If you can see the Chart toolbar, click on the Format Plot Area tool as shown in Figure 6.11. Otherwise, use the menu command Format|Selected Plot Area to bring up the Format Plot Area dialog box as shown in Figure 6.12.

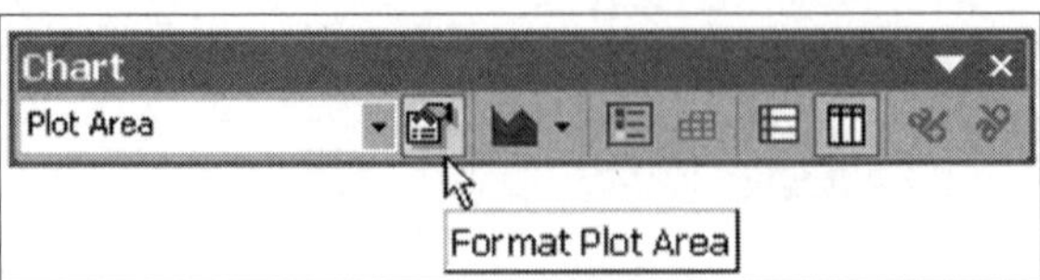

Figure 6.11

(b) Click on the *None* radio button within the *Area* region of the dialog box.

(c) Before clicking the *OK* button you may wish to explore the options that appear when you use the *Fill Effects* button.

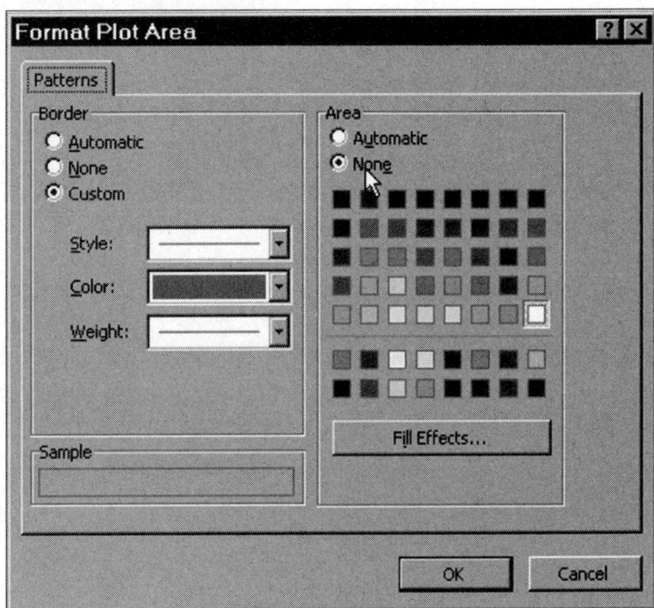

Figure 6.12

2. Format Data Series

Perhaps you would prefer hollow circles rather than solid diamond shapes used for the markers or you may wish to have no markers. Maybe you would like to have a dotted line or you may wish to have no line joining the markers.

(a) To format a data series we could begin by clicking on the line or a marker of the data series and proceeding as before. Instead, we will right click on the line or one of the markers to activate a popup menu. The first item should be *Format Data Series*. If it is *Format Data Point*, you have mistakenly clicked twice and will need to click elsewhere and start again.

(b) The Patterns tab of the Format Data Series dialog box is shown in Figure 6.13. There is one area for the line and another for the markers. Click the radio button next to *None* to remove either the line or the marker. Use the pull down arrow next to *Style* to change the type of line or the shape of the markers. Similarly the colour used for either the line or the markers may be changed. To get hollow markers make the background colour white. The *Size* value sets the marker size.

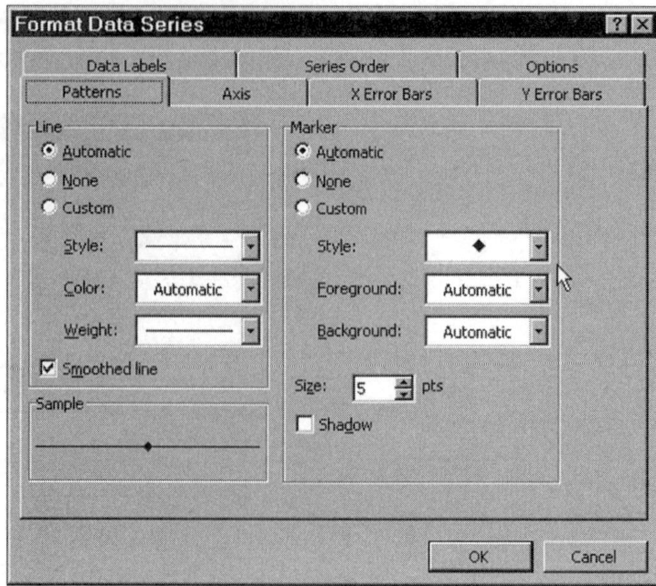

Figure 6.13

3. Format Axis: change scale

We noted earlier how the maximum value of the x-axis will depend on the size you have made your chart. Now we wish to fix the maximum value.

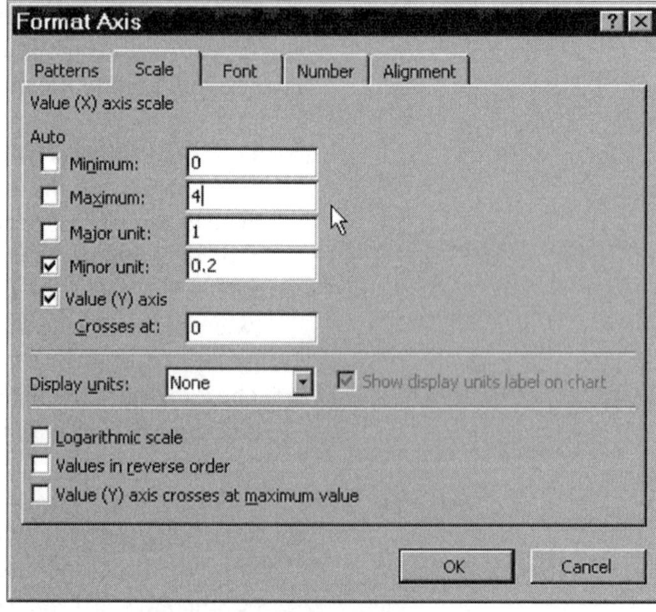

Figure 6.14

(a) This time we will go straight to the popup menu by right clicking on the x-axis. Select *Format Axis* and open the *Scale* tab to get the dialog box – Figure 6.14.

(b) Replace the value (5 or 6) in the *Maximum* box with 4. Note how the *Auto* box becomes unchecked.

(c) Does the x-axis of your chart display the number 1, 2, 3, ... or does it display 2, 4, etc.? If the latter is the case, change the value in *Major unit* from 2 to 1. Click the *OK* button.

4. Chart Options

We can add/subtract or change many features of the chart. Here are two ways to change the Chart and axes titles.

(a) For the first method, right click anywhere on the chart to open the popup menu. Locate the item *Chart Options*. This opens the same dialog box that we saw in Step 3 of the Chart Wizard. It will resemble Figure 6.8 except that the box title will read simply *Chart Options.* Make any changes you wish here and close the dialog box.

(b) For the second method, select the chart title by clicking on it. A box with handles appears. When the mouse pointer touches the border, the point changes to an open arrow. Now you can drag the title to a new location on the chart. Note the slight inconsistency – normally when you are dragging an object the mouse pointer turns into a four-headed arrow but in this case it remains as a large open arrow.

With the box still showing, start typing a new title (perhaps, *An Example of an XY Chart*). At first you may think nothing is happening but look in the Formula box. Press the formula bar's green check mark when you have completed the title.

(c) We next try a modification of the second method. Click on the x-axis title to bring up the fill handles, then click inside their box and start typing. Give the axis the title *Speed km/hr*. Now repeat the process but use the title *Speed km h-1* and then select the last two characters. Use the menu command Format|Selected Axis Title to turn these characters into superscripts thus making the title *Speed km* h^{-1}.

Using what was learned in the last step, you should now be able to change the font for any one of the titles.

5. Linking a Chart Title to a Cell Value

There are times when it is convenient to have a chart or an axis title linked to a cell on a worksheet. You may wish the title to be dynamic – change as the cell value changes. Alternatively, you may wish to use symbols as we did in Exercise 13 of Chapter 2; these are not possible in regular titles.

(a) Type some text (e.g. My XY Chart) in A9.

(b) Click on the existing chart title to bring up the fill handles. In the Formula box type an equal sign (=). Now use the mouse to point at cell A9. The Formula box displays =Sheet1!A9. Click the green check mark on the formula bar to complete the process. Experiment with changing the value in A9 and observe that the chart title changes. To use this linking method, the chart must have some title beforehand. Unfortunately, any formatting done in A9 is not reflected in the title.

6. Extending a Data Series Range

We will see how to include some new values into a data series on an existing chart.

(a) Quickly make a new chart from the data in A1:B6.

(b) Add new data values 5 and 13 into A7 and B7, respectively.

(c) Click on the chart. Note the coloured range finder lines around the x-values A2:A6 and the y-values B1:B6. The cell B1 is also outlined since this is used to name the data series. Pull down the fill handle (lower right corner) of the x-range and of the y-range. Sometimes you are lucky and pulling the x-range also pulls down the y-range. Observe the chart to see that it has been extended.

The *Source Data* dialog (Figure 6.13) may also be used to change the data series range. It is the appropriate tool to use when making more extensive changes.

Exercise 3: Line Chart with Two Data Series

In this exercise you are asked to plot the data in the table in Figure 6.15 and create a chart similar to that in the figure. You learnt most of the required techniques in Exercise 1.

(a) Enter the data on Sheet2 of the CHAP6.XLS workbook. Rows 2 and 15 are blank but we shall use them so that the Jan and

Dec axis labels do not overlap the plot area border. We shall see that this adds a small complication.

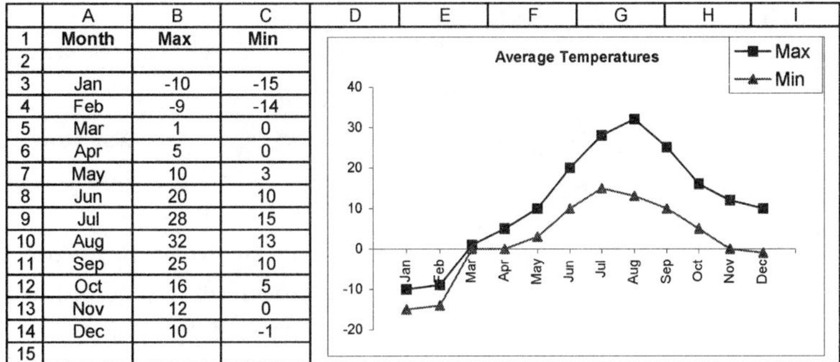

	A	B	C
1	**Month**	**Max**	**Min**
2			
3	Jan	-10	-15
4	Feb	-9	-14
5	Mar	1	0
6	Apr	5	0
7	May	10	3
8	Jun	20	10
9	Jul	28	15
10	Aug	32	13
11	Sep	25	10
12	Oct	16	5
13	Nov	12	0
14	Dec	10	-1
15			

Figure 6.15

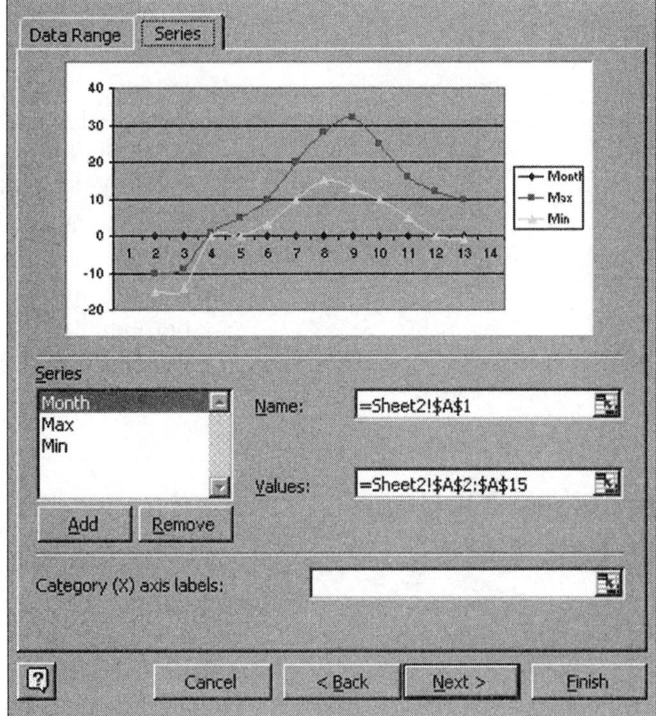

Figure 6.16

(b) Select A1:C15, start the Chart Wizard and, in Step 1, choose a Line chart with markers and lines. We do not have the option of smooth lines when initially making a Line chart. We shall address this in (e) below. Press the *Press and Hold to View Sample* button. To our surprise and annoyance, there are three data series. The *Month* series runs along the *x*-axis. The blank

‡ We could avoid this problem by putting dummy text into A2 and A15 while creating the chart and deleting it when the chart was ready. However, you will run into this problem when making a Line chart when the *x*-category values are numeric, so it is useful to learn how to solve the problem another way.

cell‡ in A2 has confused Microsoft Excel into thinking we have numeric data in column A. We solve this in Step 2 so press *Next* to continue.

(c) Open the *Series* tab of Step 2 – see Figure 6.16. Select the *Month* series and remove it. Locate the text box labelled *Category (X) axis labels* and enter Sheet2!$A$2:$A$15. Pointing is the easiest way to do this.

(d) Move on to Step 3. Add the title, remove the gridlines and position the legend at the top. The legend can be dragged into position later. In Step 4 have the chart placed on the worksheet.

(e) There is some tidying up to do:

- (i) Position the chart where you want it and resize it.
- (ii) Format the plot area to remove the grey fill and the border.
- (iii) Enlarge the plot area, if needed.
- (iv) Open the Format Data Series dialog for each line and uncheck the smoothing option–inappropriate here.
- (v) Drag the legend to the top right corner.
- (vi) Format the two axes. You may need to decrease the font size of the labels. Excel seems to want to make presentation charts with large fonts whereas we mainly use charts in reports and need smaller fonts. You will also need to adjust the alignment of the *x*-axis labels.

Exercise 4: XY Chart with Two *Y*-axes

In this exercise we graph the data in Figure 6.17. Note the range of values for *Temp* is about 10 to 25, while for *Light* we have 0 to 120. The first chart produced will resemble the upper chart of the figure. We then modify it to show the two *y*-axes.

(a) On Sheet3 of CHAP6.XLS, enter the data and create an XY chart similar to the upper one in the figure.

(b) Right click on the *Light* data line. Select *Format Data Series* from the popup menu. In the dialog box, select the *Axis* tab and click on the *Secondary Axis* radio button. Click the *OK* button.

(c) Now we have a chart with two *y*-axes but it needs more work. Right click on the plot area and choose *Clear* from the popup menu to remove both the grey fill and the plot area border.

(d) Again right click on the *Light* data line and use the *Patterns* tab to make the markers open squares so they stand out from the other data series.

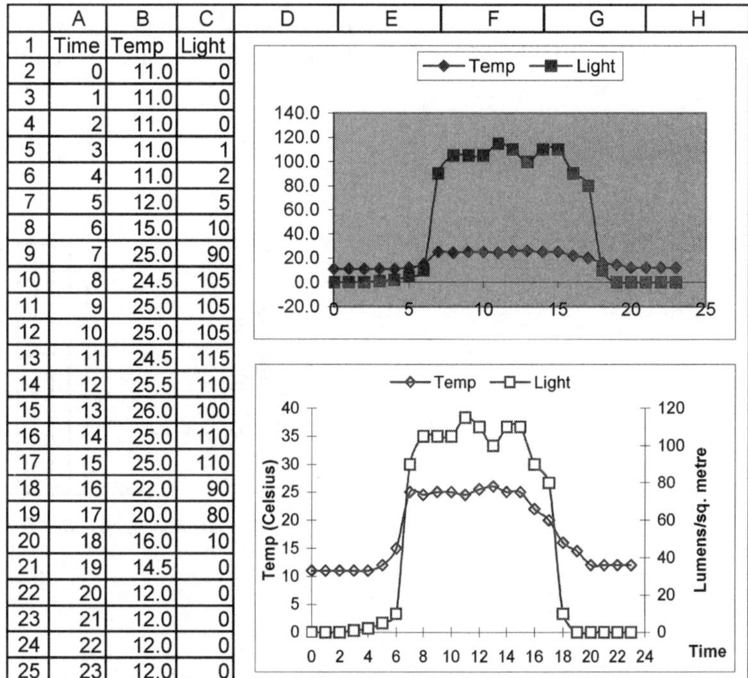

	A	B	C
1	Time	Temp	Light
2	0	11.0	0
3	1	11.0	0
4	2	11.0	0
5	3	11.0	1
6	4	11.0	2
7	5	12.0	5
8	6	15.0	10
9	7	25.0	90
10	8	24.5	105
11	9	25.0	105
12	10	25.0	105
13	11	24.5	115
14	12	25.5	110
15	13	26.0	100
16	14	25.0	110
17	15	25.0	110
18	16	22.0	90
19	17	20.0	80
20	18	16.0	10
21	19	14.5	0
22	20	12.0	0
23	21	12.0	0
24	22	12.0	0
25	23	12.0	0

Figure 6.17

(e) Right click on the *x*-axis. On the *Scale* tab, set the minimum to 0 and the maximum to 24. Set the major unit to 2. On the *Font* tab change the font size if necessary.

(f) Format the left *y*-axis so that the minimum is 0 and the maximum is 40. When a chart is made, the axis labels take on the format of the values on the worksheet but this may be changed. Format the labels to display zero decimals.

(g) Format the right *y*-axis so that the minimum is 0 and the maximum is 140. Adjust the font size if needed.

(h) Add appropriate titles to the *x*-axis and the two *y*-axes. Move the *x*-axis title.

(i) Maximize the plot area. The *y*-axis titles can be moved if required. The size of the legend box may be changed and the border removed – right click and select *Format Legend*.

Exercise 5: Combination Chart

Occasionally, you may wish to have a chart in which one data series is displayed as columns and another as a line. Such a chart may be made by going to the *Custom Type* tab in Step 1 of the Chart Wizard and selecting one of the Line and Column types. However, this does not give you control over which data series becomes the line and which the columns.

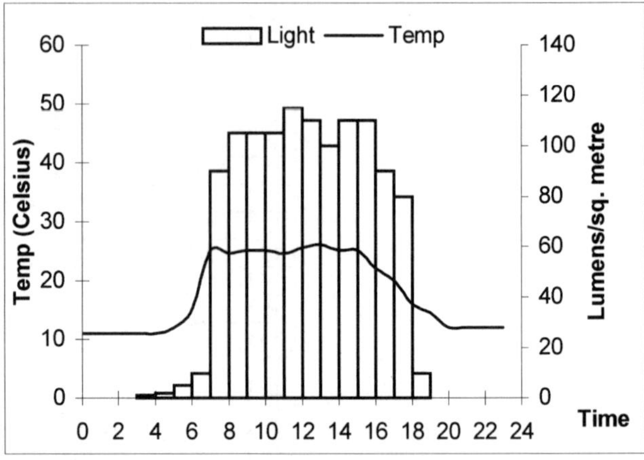

Figure 6.18

(a) Return to Sheet 3 where you completed the previous exercise. Make a copy of the second chart using Copy and Paste.

(b) Right click on the *Light* data series and from the popup menu select the *Chart Type* item. Since we currently have a data series selected, the change we are about to make will affect only that series, not the entire chart. From the options presented (the dialog box will resemble Step 1 of the Chart Wizard – Figure 6.8) select *Column.*

(c) The *Light* data series is now a set of columns. Open up the *Format Data Series* dialog box for this series and go to the *Options* tab. Change the value of the *Gap* parameter to zero. Your chart will now resemble that in Figure 6.18.

Exercise 6: Chart with Error Bars

In this exercise we explore the options for adding error bars. Although we will work only with the *y*-axis, error bars may be added to either or both axes.

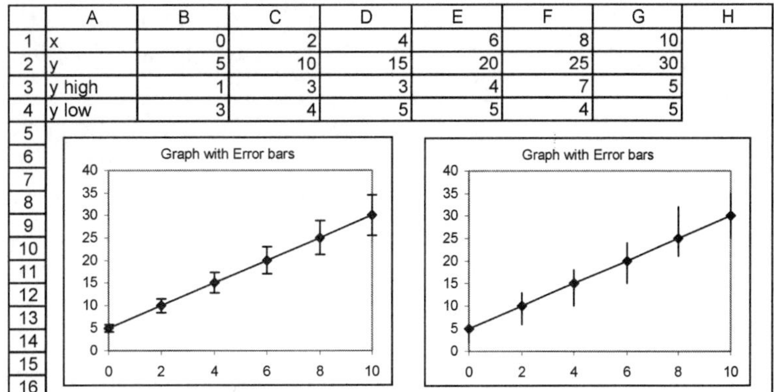

	A	B	C	D	E	F	G	H
1	x	0	2	4	6	8	10	
2	y	5	10	15	20	25	30	
3	y high	1	3	3	4	7	5	
4	y low	3	4	5	5	4	5	

Figure 6.19

(a) Enter the required data on Sheet4 of CHAP6.XLS and, using the data in rows 1 and 2 of Figure 6.19, construct an XY chart similar to that on the left of the figure but without the error bars.

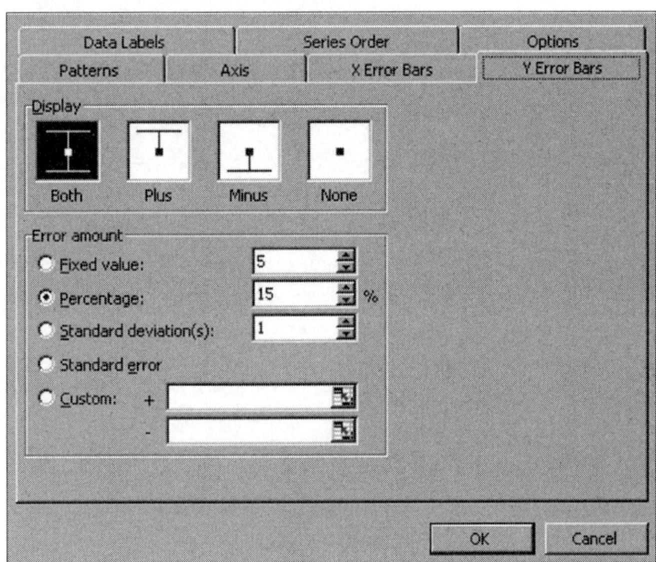

Figure 6.20

(b) Open the *Format Data Series* dialog box and move to the *Y Error Bars* tab – see Figure 6.20. Note the various options. Error bars have a positive and a negative component and we may display either, both or none. For the size of the error bars we have many options. For this exercise, choose the first *Display* type (error bars with both plus and minus components) and under *Error Amount* select *Percentage* with the value

15%. When you click *OK*, the chart will now have error bars as shown in the left of the figure.

(c) Use Copy and Paste to make a duplicate of the chart. Again open the *Y Error Bars* tab. This time use *Custom* for the *Error Amount*. In the plus text box enter =Sheet4!$B$3:$G$3 and in the minus area =Sheet4!$B$4:$G$4 – the easiest way is by pointing. The resulting chart (right side of the figure) has error bars whose size is set by the values in rows 3 and 4. Right click on an error bar and open the *Format Error Bar* dialog box. Discover how to make the error bars with and without markers.

Exercise 7: Changing Axis Crossings

There are times when we wish to change where one axis crosses the other. In Figure 6.21 we have plotted a series of negative *y*-values. The default for a Microsoft Excel chart is for the *x*-axis to cross the *y*-axis as shown in the left-hand chart. Most of us would prefer the right-hand chart.

(a) On Sheet5 of CHAP6.XLS, enter the data and use Chart Wizard to create the left-hand XY chart.

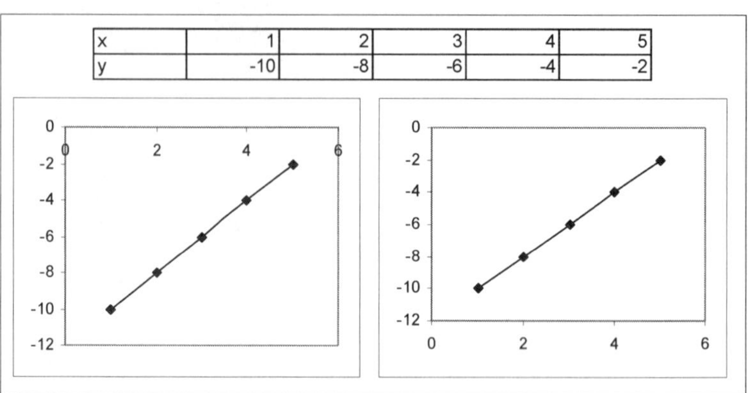

x	1	2	3	4	5
y	-10	-8	-6	-4	-2

Figure 6.21

(b) Make a copy of the chart. We wish to change where the *y*-axis crosses the *x*-axis. Open the *Format Axis* dialog for the *y*-axis and move to the *Scale* tab. The fifth box down is *Value (X) Axis Crossing*. Change the value from 0 to −12 and click OK.

Exercise 8: Blank Cells in a Data Series

We may find that our data has a missing *y*-value. How do we wish our chart to appear? The data in Figure 6.22 represents a survey of black ducks. An early morning class prevented us from doing the count on Thursday.

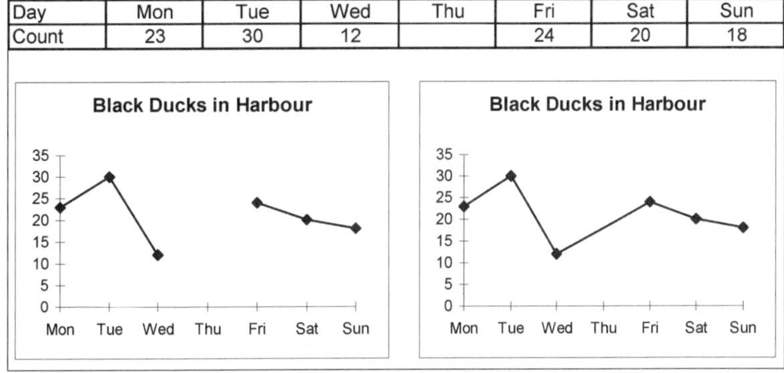

Day	Mon	Tue	Wed	Thu	Fri	Sat	Sun
Count	23	30	12		24	20	18

Figure 6.22

> There is also an option in the Tools menu for making charts with empty cells plot in this manner. The chart must be selected before opening the dialog box.

(a) Construct the Line chart shown on the left in the figure on Sheet6. Note the gap produced by the missing data.

(b) If we wish to join the Wednesday and Friday data, as in the right-hand chart, we need to add the formula =NA() (it will display as #N/A) to the empty Thursday data cell.

Exercise 9: Selecting Non-adjacent Data

In the exercises above, the *x* and *y*-values have been in neighbouring columns or rows. Often the data we wish to plot is not adjacent on the worksheet. There are two ways to select such data.

(a) On Sheet7 of CHAP6.XLS, enter the values1, 2, 3, 4 in A1:A4 and 2, 4, 6, 8 in C1:C4.

(b) The first way of selecting the two ranges is:

- (i) Select the first range A1:A4.
- (ii) While holding down Ctrl, select the second range (C1:C4).

(c) The second way looks more complicated but is often the best method.

- (i) From the Edit menu choose the Go To command.
- (ii) In the *Reference* box type the two ranges, separated by a comma, e.g. A1:A4, C1:C4. Click the OK button.

Exercise 10: A Chart with Two *X*-Ranges

(a) On Sheet8 of CHAP6.XLS, enter the text and values shown in Figure 6.23. Make an XY chart using the data in A3:B6 in the usual way. Note the secondary tick marks on the *x*-axis.

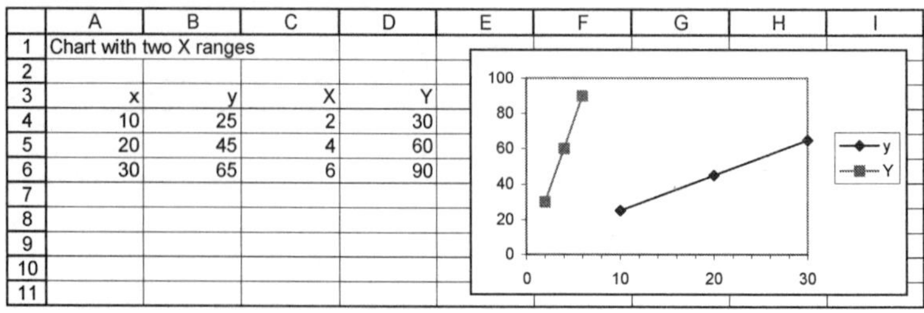

	A	B	C	D
1	Chart with two X ranges			
2				
3	x	y	X	Y
4	10	25	2	30
5	20	45	4	60
6	30	65	6	90
7				
8				
9				
10				
11				

Figure 6.23

(b) Select the range C3:D6 and click the Copy button.

(c) Click on the chart to select it. Use the menu command Edit|Paste Special. The two radio buttons *Add Cells as New Series* and *Values of Y in Columns* will have been selected. Similarly, there will be a check mark in the box for *Series Names in First Row*. You must ensure that there is a check mark in the box *Categories (X Values) in First Column*. Click the OK button. Your chart should resemble that in the figure. See also Problem 2 for another way to add a data series.

Exercise 11: A Bar Chart with a Difference

The techniques learned in this exercise may be applied to make a variety of Bar charts including critical-path (Gantt) charts. We will make the chart shown in Figure 6.24.

	A	B	C	D
1	**Acid-base Indicators**			
2				
3	Indicator	Colour-change	Low pH	ΔpH
4	Thymol blue (1)	red-yellow	1.2	1.6
5	Methyl orange	orange-yellow	3.1	1.3
6	Methyl red	red-yellow	4.2	2.1
7	Thymol blue (2)	yellow-blue	8.0	1.6
8	Phenolphthalein	colourless-pink	8.3	1.7
9				
10				
11				
12				
13				

Phenolphthalein
colourless-pink
Thymol blue (2)
yellow-blue
Methyl red
red-yellow
Methyl orange
orange-yellow
Thymol blue (1)
red-yellow
0 2 4 6 8 10

Figure 6.24

(a) Enter the data in A1:C8 on Sheet9 of CHAP6.XLS. Select A4:A8 and C4:D8 (see Exercise 10) and make a stacked Bar chart by using the second subtype in step one of Chart Wizard.

(b) Format Series 1 to make it invisible: no border, no area colour. Format Series 2 to have no area colour.

(c) Select Series 2 and open the *Format Data Series* dialog box. Go to the *Data Labels* tab and choose the *Show Values* option. Close the dialog box.

(d) Click once on the data label for Thymol blue (1); now click it again. It should be surrounded by a box. In the formula bar type an equal sign. Click on cell B4 to make the formula =Sheet9!$B$4 and complete the formula using the formula bar's green check mark (✅). If you cannot see cell B4 when the chart is activated, you will need to type this formula. The textual data labels will not fit in the data bar, so drag the data label box to one side.

(e) Repeat step (d) for the other data points.

This type of chart can be made by selecting *Floating Bars* from the *Custom Charts* tab in Step 1 of the Chart Wizard. However, the results need so much formatting that one may as well start from scratch.

Exercise 12: Displaying Units

Excel 2000 has a feature which is useful when the values in a chart are large. It may be used with both axes. Unfortunately, it is not applicable to small numbers.

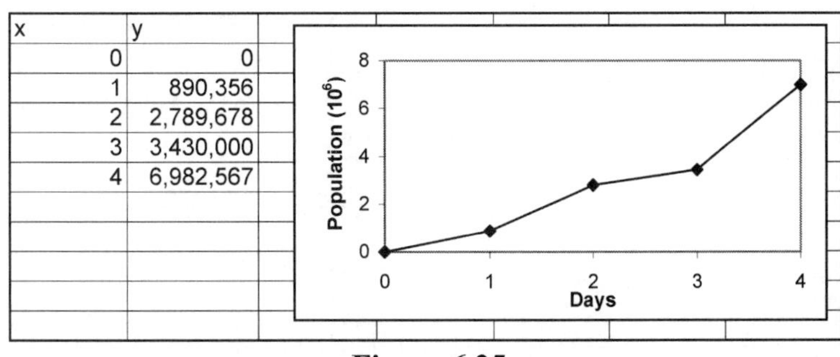

x	y
0	0
1	890,356
2	2,789,678
3	3,430,000
4	6,982,567

Figure 6.25

(a) On Sheet10, enter the data shown in Figure 6.25 and construct an XY chart.

(b) Open the Format Axis dialog box for the y-axis and go to the Scale tab. In the *Display units* scroll box, select *Millions.* Experiment with the *Show display units label on chart* option.

Exercise 13: Setting the Default Chart Type

Perhaps you would like the Chart Wizard always to open with a certain chart type selected.

(a) Open Sheet1 of CHAP6.XLS and select the chart. Use the menu command Chart|Chart Type. The dialog box that appears will resemble Step 1 of the Chart Wizard (Figure 6.6) except that near the bottom there will be a button labelled *Set as chart default*. We will opt for an XY chart with markers joined by a smoothed line as the chart type. Now click the aforementioned button and reply Yes to the next dialog box.

(b) Now select A1:B6 and click on the Chart Wizard tool. Making no changes to the chart type, click the finish button. You will have an XY chart with markers joined by a smoothed line.

This is not totally satisfactory. Your chart has gridlines and a grey plot area. That will be corrected in the next part of the exercise.

(c) On Sheet11 of CHAP6.XLS enter the data shown in A1:B6 of Figure 6.26 and construct the chart shown at the top. Note that the only formatting required is to remove the fill from the plot area as we did in (1) of Exercise 2.

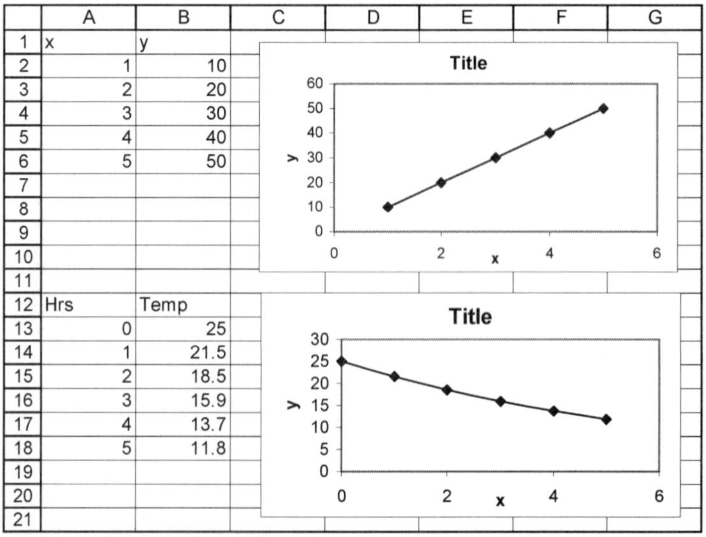

	A	B
1	x	y
2	1	10
3	2	20
4	3	30
5	4	40
6	5	50
7		
8		
9		
10		
11		
12	Hrs	Temp
13	0	25
14	1	21.5
15	2	18.5
16	3	15.9
17	4	13.7
18	5	11.8
19		
20		
21		

Figure 6.26

(d) With the chart selected, use the menu command Chart|Chart Type and go to the *Custom Types* tab. Click on the *User-defined* radio button and then on the *Add* button to open a dialog box similar to Figure 6.27. Enter a name and a description in the appropriate text boxes. Click the OK button. Click on the *Set as chart default* button before closing the dialog box.

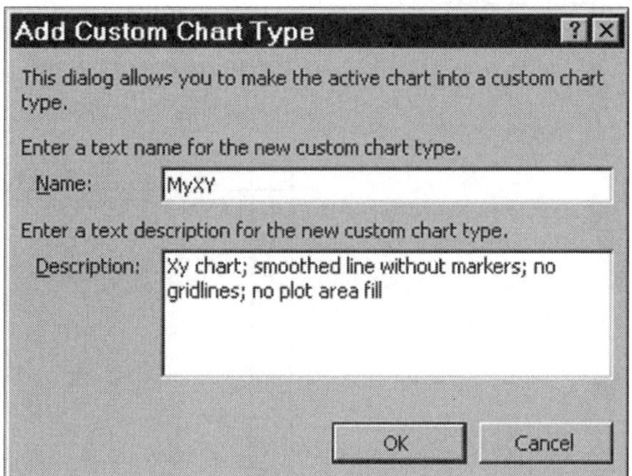

Figure 6.27

(e) To test the result, enter the data shown in A12:B18 of Figure 6.26. With this data selected, start the Chart Wizard and immediately click the Finish button. After some resizing but no formatting, the resulting chart should resemble the lower one in the figure.

Selecting a Chart Object

We have been selecting components to be modified by clicking on them. With some charts this is not always successful. For example, in the next chapter we will add a trendline to a chart. This makes it difficult to select the underlying data series. Here are two methods to use in such cases:

An object is selected when it is outlined with solid squares and its name shows in the Name box.

(a) Activate the chart. Tap any one of the navigation keys (↓, ↑, ←, →) and observe how different components are selected in turn. When the required component is selected, use the Format menu item to open the appropriate formatting dialog.

Figure 6.28

(b) Use View|Toolbars and put a check mark beside *Chart*. When the chart is selected, the Chart tool has a drop down list of components – Figure 6.28. Make your choice and use the Format item on the menu to open a dialog box. The tool also has a variety of other uses; the reader is invited to experiment.

Too Much Data

Sometimes we have too many data points to make an attractive chart. If you had 100 points, the chart would be unattractive with the markers crowded together. For a simple way to solve this, we will start with just 10 data points – see Figure 6.29.

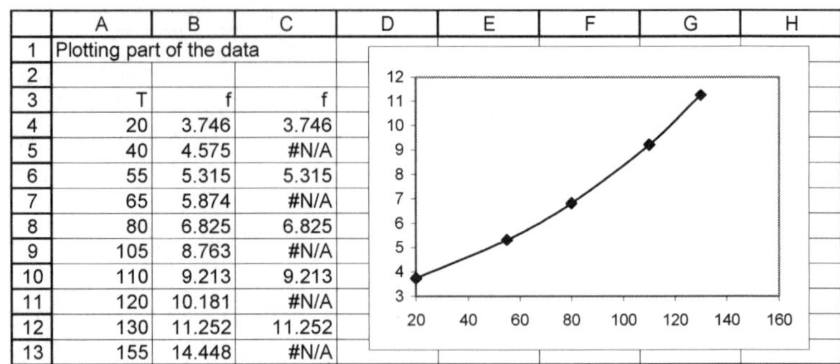

	A	B	C
1	Plotting part of the data		
2			
3	T	f	f
4	20	3.746	3.746
5	40	4.575	#N/A
6	55	5.315	5.315
7	65	5.874	#N/A
8	80	6.825	6.825
9	105	8.763	#N/A
10	110	9.213	9.213
11	120	10.181	#N/A
12	130	11.252	11.252
13	155	14.448	#N/A

Figure 6.29

The 'raw' data is in columns A and B. The formula in C4 is =IF(MOD(ROW(),2)=0, B4, NA()). This is copied down the column and results in every second value in column B being displayed in column C. Selecting columns A and C by the method explored in Exercise 9, we make the chart shown.

Of course, if we wished to plot every tenth row of a larger data set then =IF(MOD(ROW(),10)=0, B4, NA()) would do the trick but it might not capture the first point. If the first data point is in row 3, then we could use =IF(MOD(ROW(),10)–3=0, B4, NA()).

Another useful trick is to plot both the raw and the filtered data; the former as just a line and the latter as just markers.

Dynamic Charts

If you collect data over a prolonged period you may wish to be able to add data to a worksheet and have a chart automatically updated to show the new information. Since this is more of a business application, we will not undertake an exercise on this topic but interested readers are directed to the author's web site for details on how to do this.

Printing a Chart

If an embedded chart (a chart on a worksheet) is selected when the print command is issued, the printout consists of just the chart. It is expanded to fill the page. Conversely, there may be times when you wish to print a worksheet without the chart. Right click on the chart and select Format Chart Area. On the Properties tab, remove the check mark from the Print Object box.

Problems

1. Create the XY chart shown to the left in the figure below. Make the following changes so that it becomes similar to the right-hand chart.

   (a) Alter the *y*-axis minimum and maximum values.
   (b) Change the type of major tick marks and add minor ticks to the *x*-axis.
   (c) Change the style of the line and markers, enlarge the markers.
   (d) Change the text, position and font of the title.

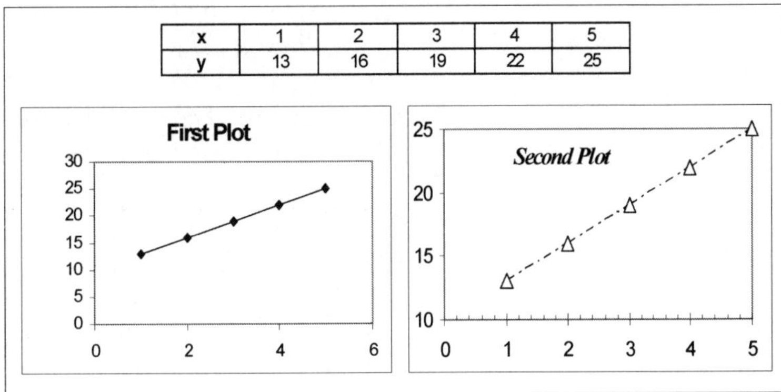

x	1	2	3	4	5
y	13	16	19	22	25

2.* Your task is to add a new data series to an existing chart. We did this in Exercise 8 using Edit|Paste Special. There is another way. Open CHAP6.XLS on Sheet1 and add new data in column C. In C1 enter Z for name of the data, and in C2:C6 enter 2, 4, 6, 8, 10. Right click on the chart, select the *Source Data* item. What must you do to complete the task?

3.* You need to take to a work site a chart as shown in the figure below. The secondary axes are needed to help you draw horizontal and vertical lines on the chart. What steps are needed to make the chart?

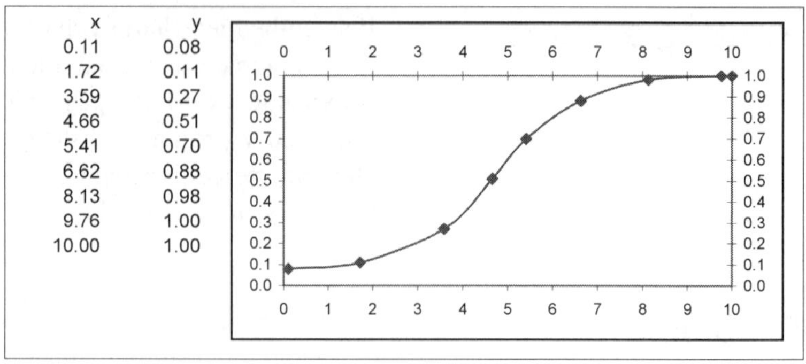

x	y
0.11	0.08
1.72	0.11
3.59	0.27
4.66	0.51
5.41	0.70
6.62	0.88
8.13	0.98
9.76	1.00
10.00	1.00

4. Construct the chart shown below. The data in A1:B10 is used to chart the parabola $y = 4x^2$. When the value in E18 is varied, the two straight lines and the marker move to new positions. The formulas in A12, A14 and A16 are all =E18.

   Use either of the methods you have learned to add new series to a chart to add the two lines and the marker. Format the lines appropriately and format the marker to display the *x*- and *y*-values – look for the Data Labels tab in the Format Data Series dialog.

   Use the Data|Validation command to ensure the user keeps E18 within the range ±4.0.

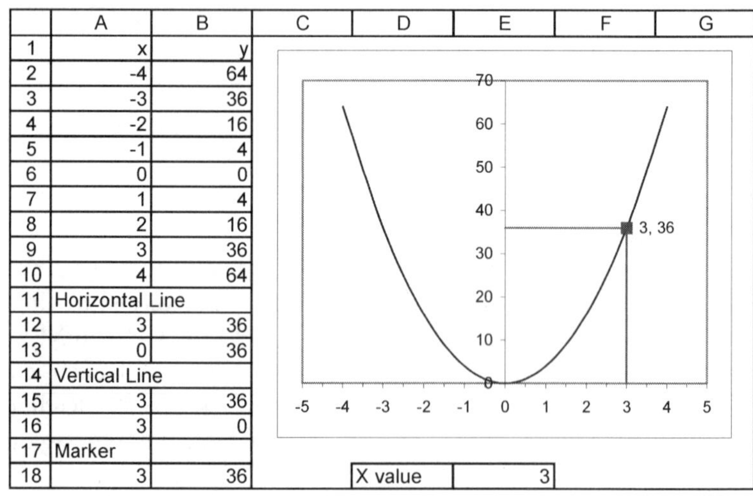

	A	B
1	x	y
2	-4	64
3	-3	36
4	-2	16
5	-1	4
6	0	0
7	1	4
8	2	16
9	3	36
10	4	64
11	Horizontal Line	
12	3	36
13	0	36
14	Vertical Line	
15	3	36
16	3	0
17	Marker	
18	3	36

7
Curve Fitting

Concepts

There are many occasions where we plot data to obtain an experimental result. We will consider some simple experiments in which one variable (X, the independent variable) is varied by the experimenter and the dependent variable Y is measured. Frequently we arrange things such that $Y = mX + b$, where m and/or b are the quantities we wish to determine. By using a linear relationship we are able to plot X against Y to get a straight line. The slope and the intercept of the graph give us the values of m and b, respectively.

Example A
If a body has an initial velocity v_0 and an acceleration a, then its velocity v at time t is given by:

$$v = v_0 + at \qquad (7.1)$$

This may be rearranged as $v = at + v_0$ which has the form $y = mx + b$. Hence, a plot of velocity v on the y-axis against time t on the x-axis will have a slope equal to the acceleration a and an intercept equal to the initial velocity v_0.

Example B
Consider a body with initial velocity v_0 accelerating at a rate a. The distance d that it travels in time t is given by:

$$d = v_0t + \tfrac{1}{2}at^2 \qquad (7.2)$$

This is not a linear equation. But if we can arrange the experiment such that v_0 is zero, then Equation 7.2 becomes

$$d = \tfrac{1}{2}at^2 \qquad (7.3)$$

Comparing this to $y = mx + b$, we see that plotting d against t^2 (rather that t) will give a linear graph. The slope of this line will have the value $\tfrac{1}{2}a$ from which we may readily evaluate a.

Example C
A simple model for the growth of bacteria predicts that if the initial population is C, the population N at time t will be given by Equation 7.4 in which B is the birth-rate:

$$N = C \exp(Bt) \qquad (7.4)$$

This is not a linear relationship. However, with some simple mathematical manipulation we can produce a linear form of the equation – we can linearize it. Take the logarithm of both sides to get Equation 7.5 which rearranges to 7.6.

$$\ln(N) = \ln(C) + Bt \quad (7.5)$$

$$\ln(N) = Bt + \ln(C) \quad (7.6)$$

Comparing this to $y = mx + b$ shows that plotting $\ln(N)$ against t will give a linear graph with a slope of B and an intercept of C.

Least squares fit

In each of these examples we see that we need to find the slope of a linear graph. Remember that experimental errors will mean that the data does not exactly fit a straight line. Before computers we would plot the data and then, with a ruler, draw a line that went as close as possible to the points. It is interesting to see the spread of values for the slope obtained in this way by a group of people using the same set of data values.

There are more precise ways of determining the line of best fit for linear data. The experimental data consists of pairs of x and y values. We write the equation of the line of best fit as $\hat{y} = mx + b$, where $\hat{y}$ (read as 'y hat') is the *predicted value*. The *least squares criterion* requires that we adjust the constants m and b such that the quantity $\Sigma(y - \hat{y})^2$ is as small as possible. The required calculations are tedious. Fortunately, spreadsheets relieve us of the problem. We will explore some of the Microsoft Excel features relating to curve fitting in this chapter and again in Chapters 10 and 14. We will start with exercises using the examples above.

Exercise 1: Finding the Slope and Intercept

The data in Table A of Figure 7.1 shows the velocity of an accelerating body at various times – see Example A above. We will use the SLOPE and INTERCEPT functions to find the acceleration and initial velocity.

	A	B	C	D	E	F	G	H	I	J
1			Table A							
2	Time	2	4	6	8	10			Best Fit Parameters	
3	Velocity	22	42	62	80	100		slope	acceleration	9.7
4	Best Fit	22.4	41.8	61.2	80.6	100.0		intercept	initial velocity	3

Figure 7.1

(a) Open a new workbook and enter the text and data in the range A1:I4 to Sheet1.

fx Insert Function tool

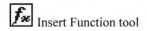

(b) In J3 enter =SLOPE(B3:F3, B2:F2). This will return the slope of the line of best fit for the data. Remember that in addition to simply typing this formula we can use the Insert Function dialog which may be called (i) using the Insert Function tool or (ii) by typing the start of the formula =SLOPE and using Ctrl+A to bring up the Function Argument dialog box.
Note the syntax of the function is:
=SLOPE(***known_Y_values***, *known_X_values*).

Take care to remember this, since it seems 'backwards' to most scientists and engineers who are accustomed to listing x-values before y-values.

(c) In J4 enter =INTERCEPT(B3:F3, B2:F2). This will return the value of the intercept of the line of best fit. The syntax is INTERCEPT(***known_Y_values***, *known_X_values*).

(d) Save the workbook as CHAP7.XLS.

Knowing the m and b values for the best fit line $\hat{y} = mx + b$, we could use the formula =$J$2*B2+$J$3 in cell B4 and copy it to C4:F4. Alternatively, we could use the TREND function to place the y values for the best fit in B4:F4. We might then plot A2:F4 showing the experimental data (B3:F3) with markers and no connecting line, and the best fit data (B4:F4) with a line and no markers. The reader is encouraged to experiment with both methods. But there is a quicker way as we will see in the next exercise.

Exercise 2: Adding the Trendline to a Chart

Microsoft Excel has a feature for plotting the line of best fit on an XY chart. This is called the *trendline*. In this exercise we will see how to add a trendline and how to extend it. In the subsequent exercise we will learn how to display on the chart the equation of this line of best fit.

(a) On Sheet1 of CHAP7.XLS construct an XY chart of the data in the range B2:F3. In Step 1 of the Chart Wizard select the first XY subtype which shows the data plotted with markers but no joining line.

(b) Right click on any marker and select *Insert Trendline* from the resulting menu. A dialog box is opened – see Figure 7.2. Select the thumbnail sketch of a Linear type.

(c) Open the *Option* tab of the dialog box. Make sure there are no Xs in any of the option boxes – see Figure 7.3. Click the OK button. Your graph will be similar to that in the chart shown to the left in Figure 7.4.

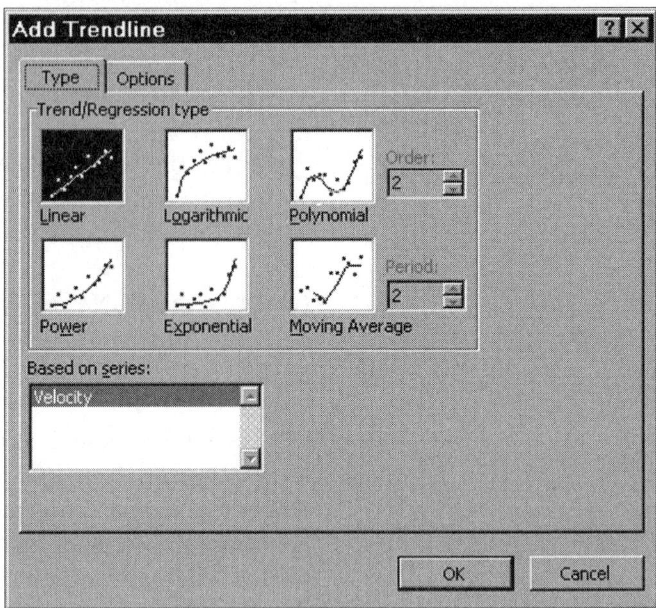

Figure 7.2

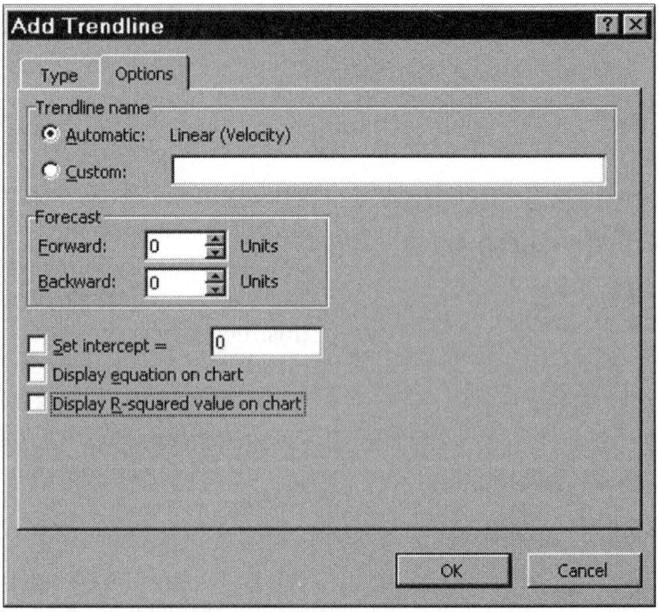

Add Trendline

Type Options

Trendline name

Automatic: Linear (Velocity)

Custom:

Forecast

Forward: 0 Units

Backward: 0 Units

Set intercept = 0

Display equation on chart

Display R-squared value on chart

OK Cancel

Figure 7.3

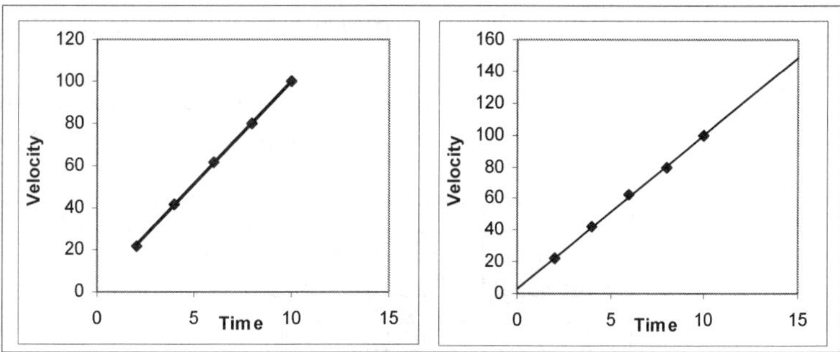

Figure 7.4

There are two features of the trendline that you may wish to change.

(d) By default, Excel draws trendlines with a thick line. Right click on the trendline, select *Format Trendline* and open the *Patterns* tab. Decrease the *weight* of the line by one.

(e) Perhaps you would prefer the line to be extended to meet the left and right sides of the plot area. Again open the *Format Trendline* dialog box and move to the *Option* tab. In the *Forecast* box, insert values of 5 and 2 in the Forward and Backward boxes, respectively. This extends the trendline from an x-value 10 to x-value 15, and from 2 to 0. After adjusting the maximum for the x-axis, your chart will resemble the right-hand chart in Figure 7.4.

Exercise 3: Adding the Trendline Equation

The data in Figure 7.5 represents the results of an experiment to measure the acceleration of a steel ball falling through a viscous liquid. At time $t = 0$ the ball is released from under the surface. The distance (in centimetres) it has moved is measured at fixed time intervals. We will assume that for the period of the measurements the ball's motion obeys the equation $d = \frac{1}{2}at^2$. If this equation is compared to the standard linear equation $y = mx + b$, we see we need to plot d against t^2. The slope of this line will be $\frac{1}{2}a$; knowing this value we may compute the acceleration. Note that the intercept of the best fit line must be zero in this instance.

Symbols and such: In Exercise 13 of Chapter 2 we learnt how to add symbols to a text entry. The squared and cubed symbols are generated with [Alt]+0178 and [Alt]+0179, respectively.

(a) On Sheet2 of the CHAP7.XLS workbook, enter the text in the range A1:C1. After typing 'Time' press [Alt]+[Enter↵], then type '(seconds)'. To achieve the superscript after typing '(sec2)', select the '2', use Format|Cells and in the dialog box click the box labelled *Superscript*.

(b) Enter the values in A2:A12 and C2:C12.

(c) In B2 enter the formula =A2^2, or, if you prefer, use =A2*A2 to give us t^2. Copy this down to B12.

	A	B	C
1	Time t (seconds)	t^2 (sec^2)	Distance (cm)
2	0.00	0.0000	0.00
3	0.05	0.0025	0.30
4	0.10	0.0100	1.25
5	0.15	0.0225	2.40
6	0.20	0.0400	4.60
7	0.25	0.0625	7.10
8	0.30	0.0900	10.00
9	0.35	0.1225	13.70
10	0.40	0.1600	18.10
11	0.45	0.2025	22.60
12	0.50	0.2500	28.00

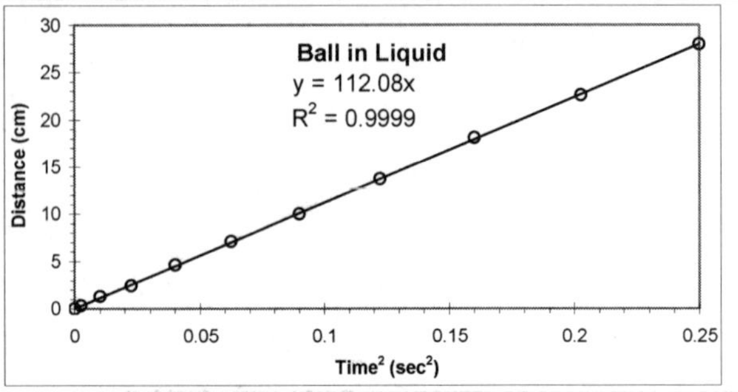

Figure 7.5

(d) Make an XY chart of the data in B1:C12 using only markers.

(e) Begin the process of adding the trendline as you did in Exercise 2 but this time on the Options tab: (i) put a ✓in the *Set intercept* box and enter the value 0 to set the intercept value, and (ii) put ✓ in the boxes labelled *Display Equation on Chart* and *Display R-squared Value on Chart*. Click on OK. Your chart should now be similar to that in Figure 7.5.

Some formatting notes: (i) After entering the x-axis title as *Time2 (sec2)*, the 2s were selected one at a time and, using the main menu Format|Selected Axis Title, a superscript font was selected. (ii) The two axes were separately modified to show minor tick marks.

The trendline equation shows the slope of the best fit line to be 112.08 cm/sec^2. We know this is equal to ½a, so the acceleration is 2.24 ms^{-2}. You may be wondering about the meaning of R^2. The short explanation is that this quantity, which is also called the *coefficient of determination*, is a measure of how well your data fits a linear equation. The closer R^2 is to unity, the better the fit. For a complete explanation of this quantity look up the topic Linear Regression in a statistics textbook.

Note that the trendline equation may be formatted and it may sometimes be advisable to do so – see Problem 5.

Exercise 4: The LINEST Function

In Exercise 1 we saw the use of the SLOPE and INTERCEPT functions. The LINEST function is somewhat more versatile. It uses the least squares method to calculate a straight line that best fits the data, and returns an array that describes the line. The syntax of this function is: LINEST(***known_Y_values***, *known_X_values, Constant, Statistics*).

If *Constant* is TRUE, or omitted, the intercept is calculated. Otherwise the intercept is set to zero and the data is fitted to $\hat{y} = mx$. When *Constant* is TRUE, the values that LINEST returns for the slope and intercept are the same as returned by the functions SLOPE and INTERCEPT. Note that using Trendline gives us a little more control. We can specify that the intercept shall have a value of, for example, 4.25.

If *Statistics* is TRUE, the function returns the value of *R*-squared and other regression statistics. We will be concerned only with R^2.

Note that LINEST returns more than one value and is, therefore, an *array function*. To use the function we must: (i) select a range for the output, (ii) type the function, and (iii) press Ctrl+Shift+Enter to complete the entry. Failure to follow these steps will result in LINEST returning only the slope.

The reader should refer to the online Help to get a list of all the statistics generated by the function. Since our data has only one set of *known_X_values*, and we wish to see the value of R^2, our output range should be a two columns by three rows range. The table below shows the arrangement of values in the output.

Slope	Intercept
Standard error in the slope	Standard error in the intercept
R-squared	Standard error in *y* estimate

In Figure 7.6, Table D gives the size of a bacteria population (N) at various times (t). In Example C in the introduction to this chapter we saw that a plot of $\ln(N)$ against t should give a linear plot of slope B, the birth-rate. We could make such a plot and insert the trendline and its equation, or we could use the SLOPE and INTERCEPT function. However, we will use the LINEST function.

	A	B	C	D	E	F	G
1	Table D	Time t	2	4	6	8	10
2		Population N	2500	6000	15000	35000	90000
3		Ln(N)	7.824046	8.699515	9.615805	10.4631	11.40756
4							
5			LINEST output			Birthrate	Initial N
6		Slope	0.44653132	6.922819	Intercept	0.446531	1015.178
7			0.00381384	0.025298			
8	R-squared		0.9997812	0.024121			

Figure 7.6

(a) On Sheet3 of the CHAP7.XLS workbook, enter the text in A1:B3 and the values in C1:G2.

(b) In C3, enter the formula =LN(C2) and copy it D3:G3.

(c) Enter the text shown in the lower half of the figure.

(d) With B6:C8 selected, type the formula =LINEST(C3:G3,C1:G1,TRUE,TRUE) and press Ctrl+Shift+Enter to complete the array formula.

The ln(*N*) values are the *known_Y_values* and Time values are the *known_X_values*. We have used *TRUE* twice so that the intercept will be calculated and *R*-squared will be displayed in the output.

We know that the slope of ln(*N*) against *t* is the birth-rate in this experiment. The intercept is ln(*C*) so the initial population *C* will be found from exp(*intercept*).

(e) In F6 enter the formula =B6 and in G6 enter the formula =EXP(C6). We see that the birth-rate is 0.45 and the initial population was about 1000.

Exercise 5: LINEST with Polynomial Data

The LINEST function may be used with more than one set of *x*-values. That is to say, one can use it with the multiple regression equation, $y = m_1x_1 + m_2x_2 + m_3x_3 + m_4x_4 + b$. The online Help uses an example to determine how the cost of an office building is related to its area, age, number of offices and number of entrances. So we may use the function to fit data to a polynomial such as $y = m_1x^4 + m_2x^3 + m_3x^2 + m_4x + b$.

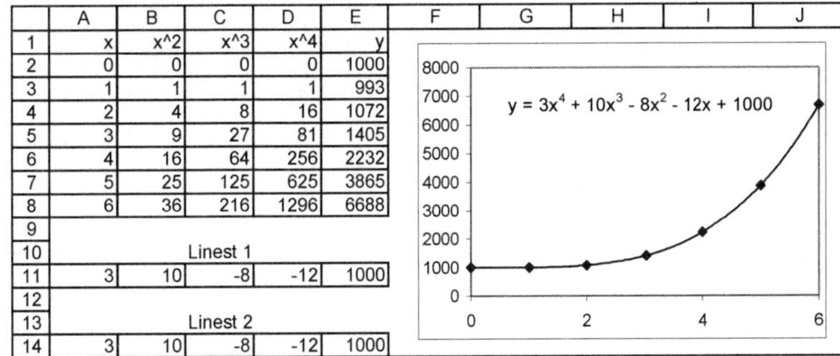

	A	B	C	D	E
1	x	x^2	x^3	x^4	y
2	0	0	0	0	1000
3	1	1	1	1	993
4	2	4	8	16	1072
5	3	9	27	81	1405
6	4	16	64	256	2232
7	5	25	125	625	3865
8	6	36	216	1296	6688
9					
10			Linest 1		
11	3	10	-8	-12	1000
12					
13			Linest 2		
14	3	10	-8	-12	1000

Figure 7.7

Suppose we have a set of (x, y) data such as that shown in columns A and E of Figure 7.7 and we wish to fit it to a quartic equation.

(a) On Sheet4 of CHAP7.XLS, enter the headers in row 1 together with the data in A2:A8 and E2:E8. Make an XY chart with only markers (see Exercise 9 of Chapter 6 to recall how to work with non-contiguous columns) and add a trendline using a fourth-order polynomial.

To have the coefficients displayed in worksheet cells we will use the LINEST equation. If we compare our problem with that in the online Help, we may be led to believe that we need columns with the x, x^2, x^3 and x^4 values. Let's try that.

(b) In B2:D2 enter =A2^2, =A2^3 and =A2^4, respectively. Copy these to row 8.

(c) Select A11:E11, enter the formula =LINEST(E2:E8,A2:D8) and press Ctrl+Shift+Enter to complete the array formula. Note that we have not used the *Constant* or the *Statistics* arguments. Omitting the first means that LINEST will compute the intercept while omitting the second means that it will not compute the statistics such as R^2. We need a range of five columns to compute the four coefficients plus the intercept. We need only one row because we are not computing the statistics.

Now we will see that the data in columns B, C and D of the table is not really necessary. We will make a two-dimensional array within the LINEST function.

(d) Select A14:E14, type =LINEST(E2:E8, A2:A8^{1,2,3,4}) and press Ctrl+Shift+Enter to complete the array formula. The *known_X_values* in this formula are computed by Excel as the values in A2:A8 raised to the first, the second, the third and the fourth power. This little trick can save some work and keep the worksheet tidy by avoiding redundant data.

Exercise 6: Non-linear Plots

We began this chapter with a discussion on linearizing equations. Our reason for doing this is mainly tradition – in the pre-computer times it was easier to draw a straight line to find the best fit. You have noticed that the Trendline dialog box gives us other options including exponential and polynomial fits. In this exercise we will see the use of an exponential fit.

(a) Open the workbook CHAP7.XLS and select Sheet3 on which Exercise 4 was completed.

(b) Select the range B1:G2 and create an XY chart with markers and no lines.

(c) Click on one of the data markers. Use the menu command Chart|Add Trendline. On the *Type* tab, select the Exponential thumbnail sketch.

(d) Go to the *Options* tab. Change the *Forecast Backward* value to 2; this will extrapolate the data to zero time. Make sure there is no X in the *Set intercept* box. Click on the next two boxes: *Display Equation on Chart* and *Display R-squared Value on Chart*. Click the *OK* button. Your chart should be similar to that in Figure 7.8. Note that the data for *slope* and *intercept* agrees with the results obtained in Exercise 4.

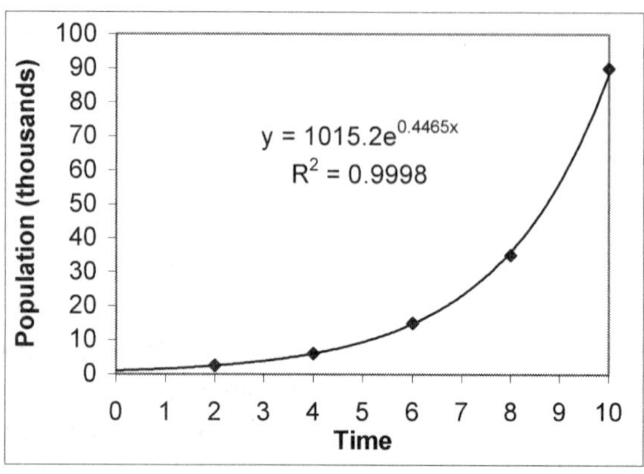

Figure 7.8

Next we will show that the same results may be obtained from the LOGEST function. This function is similar to the LINEST function but uses the logarithmic model ln(y) = xlLn(m) + ln(b) rather than the linear model. The syntax for the LOGEST function is LOGEST(***known_Y_values***, *known_X_values, Constant, Statistics*) where the arguments have the same meaning as in the LINEST function.

(e) On Sheet3, enter the text shown in Figure 7.9.

	A	B	C	D	E	F	G
11		LOGEST output				Birthrate	Initial N
12	m	1.56288164	1015.178	b		0.446531	1015.178

Figure 7.9

(f) Select B12:C12, enter the formula =LOGEST(C2:G2,C1:G1) and press Ctrl+Shift+Enter to complete the array formula. You should get the values shown in the figure.

How do we reconcile these values with those of the trendline equation in the chart? The model for LOGEST is ln(y) = xln(m) + ln(b). The latter could be written as $y = bm^x$. Compare this with the trendline equation y = bexp(kx), and we see that the b terms are the equivalent and k = ln(m).

(g) Enter =LN(B12) in F2 and =C12 in G2.

On this worksheet we have used LINEST, LOGEST and a trendline to find the parameters that mathematically describe the behaviour of the bacteria colony.

Exercise 7: Residuals

When the purpose of a regression analysis is to find which model best describes a physical process, there is often the nagging worry that some small mathematical term has been overlooked. Residual analysis can be helpful in such cases. Let y_i be the observed value and $\hat{y}_i$ the corresponding value predicted by the equation used to fit the data. The residual is defined as $e_i = y_i - \hat{y}_i$. If the prediction model is a good one, we expect the residuals to be randomly scattered about zero. If they display a pattern, we have cause to believe that a better model is possible.

In this exercise we make at a linear fit to some experimental data and examine a plot of the residuals.

(a) On Sheet5 of CHAP7.XLS enter the values shown in A1:B11 of Figure 7.10. Construct the upper chart and insert a linear trendline.

(b) Use the SLOPE and INTERCEPT function in A14 and B14. Name these cells *slope* and *intercept*, respectively.

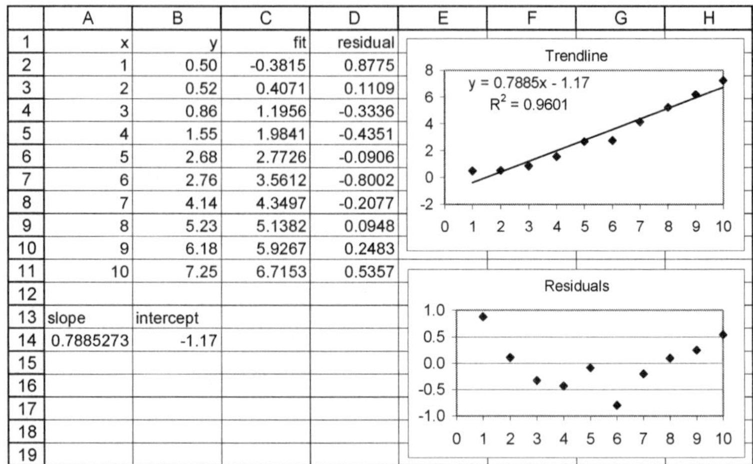

	A	B	C	D
1	x	y	fit	residual
2	1	0.50	-0.3815	0.8775
3	2	0.52	0.4071	0.1109
4	3	0.86	1.1956	-0.3336
5	4	1.55	1.9841	-0.4351
6	5	2.68	2.7726	-0.0906
7	6	2.76	3.5612	-0.8002
8	7	4.14	4.3497	-0.2077
9	8	5.23	5.1382	0.0948
10	9	6.18	5.9267	0.2483
11	10	7.25	6.7153	0.5357
12				
13	slope	intercept		
14	0.7885273	-1.17		
15				
16				
17				
18				
19				

Figure 7.10

(c) In C2 the formula =slope*A2 + intercept is used to compute the predicted values, while =B2 – C2 is used in D2 to compute the residual for this point. These are copied down to row 11.

(d) Construct a plot of the residuals (D2:D11) against the independent values (A2:A11), as shown in the lower chart.

The residual plot is not random but seems to be an approximation to a parabola. If you now carefully examine the first chart you may see that the markers do form a shallow quadratic. Right click on the trendline and change it from linear to a second-order polynomial. Use the LINEST equation in a manner similar to that in the last part of Exercise 5 to get the coefficients of the quadratic and proceed with a residual analysis for this model.

Exercise 8: Calibration Curve

A chemist makes six iron solutions with varying concentrations. He treats samples of each to convert the iron to a purple compound and measures the absorbance of 562 nm light of each sample. From this he obtains a calibration curve. When he treats samples with unknown amounts of iron in the same manner, the measured absorbance can be used to find the iron content from his plot.

(a) On Sheet6 of CHAP7.XLS, enter everything shown in A1:B9 of Figure 7.11 and construct the chart. When you add the trendline, set the intercept to zero.

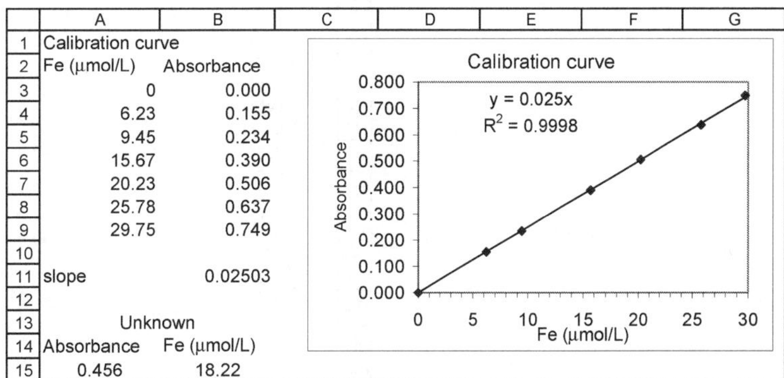

	A	B
1	Calibration curve	
2	Fe (μmol/L)	Absorbance
3	0	0.000
4	6.23	0.155
5	9.45	0.234
6	15.67	0.390
7	20.23	0.506
8	25.78	0.637
9	29.75	0.749
10		
11	slope	0.02503
12		
13	Unknown	
14	Absorbance	Fe (μmol/L)
15	0.456	18.22

Figure 7.11

(b) Compute the slope in B11 using either the SLOPE or the LINEST function.

(c) The absorbance reading is entered in A15. Since the calibration data fits the equation $y = mx$ or *Absorbance* = *slope* × *Iron content*, it follows that *Iron content* = *Absorbance/slope*. The required formula in B15 is therefore =A15/B11. Had the calibration equation been in the form $y = mx + b$, we would use in B15 a formula in the form =(*Y* – *intercept*)/*slope*.

Note that we do not really need the chart unless we wish to see a graphical representation of the calibration data. See Chapter 14 for more on this topic.

Exercise 9: Interpolation

An engineer has tested an aggregate sample, recording the percentages that pass through sieves of various sizes. Her data is shown in A2:B19 of Figure 7.12. The engineer wishes to use the worksheet to predict which size sieve will allow a specified percentage of the sample to pass through. Thus when the required percentage (*Y*) is 50, the chart shows that a sieve size (*X*) of approximately 0.16 is required. The task is to obtain this value without using a chart. Note, however, we shall use the chart to explain and confirm our method. This problem differs from the calibration curve discussed above in that there is no simple equation to fit the data, so we elect to use interpolation.

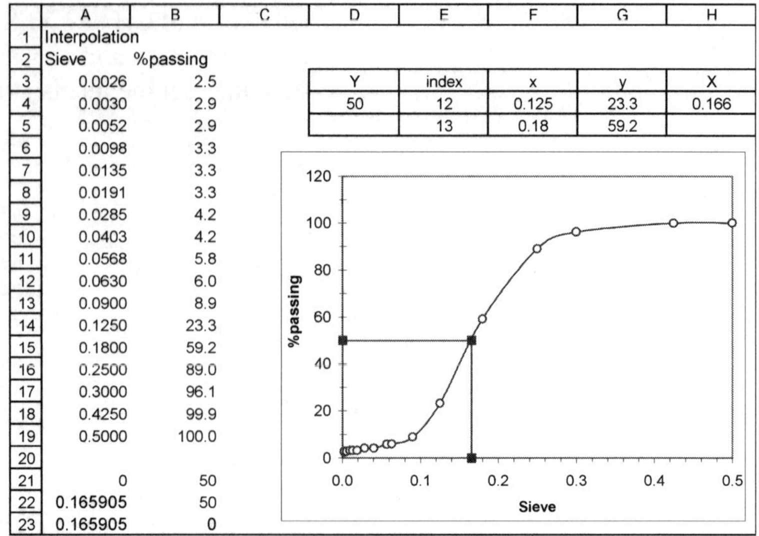

	A	B	C	D	E	F	G	H
1	Interpolation							
2	Sieve	%passing						
3	0.0026	2.5		Y	index	x	y	X
4	0.0030	2.9		50	12	0.125	23.3	0.166
5	0.0052	2.9			13	0.18	59.2	
6	0.0098	3.3						
7	0.0135	3.3						
8	0.0191	3.3						
9	0.0285	4.2						
10	0.0403	4.2						
11	0.0568	5.8						
12	0.0630	6.0						
13	0.0900	8.9						
14	0.1250	23.3						
15	0.1800	59.2						
16	0.2500	89.0						
17	0.3000	96.1						
18	0.4250	99.9						
19	0.5000	100.0						
20								
21	0	50						
22	0.165905	50						
23	0.165905	0						

Figure 7.12

(a) On Sheet7 of CHAP7.XLS, enter the text and data shown in A1:B19, and the text in D3:H3.

(b) Construct the chart using the data in A3:B19. The three points joined by straight lines will be added later.

(c) In D4, enter the value 50. The formulas in E4:H4 are:

```
E4:  =MATCH(D4,B3:B19,1)
E5:  =E4+1
F4:  =INDEX($A$3:$A$19,E4)
F5:  =INDEX($A$3:$A$19,E5)
G4:  =INDEX($B$3:$B$19,E4)
G5:  =INDEX($B$3:$B$19,E5)
H4:  =(D4-G5)*(F4-F5)/(G4-G5)+F5
```

The MATCH function in E4 locates the position in B3:B19 that has a value less than or equal to the lookup value (D4). A value of +1 is used for the third argument in the function because the values in the table are in ascending order. When $Y = 50$, the function returns position 12. Clearly, the formula in E5 merely increments this by 1. Therefore, the required X,Y pair lies between the 12th and 13th known x,y pairs.

The INDEX formulas in F4:G5 translate these positions into actual x,y pair values. Let us call these x_1, y_1 and x_2, y_2. On the chart, these are the two circles which are above and below the square marker.

If we let these two points be joined by a straight line, we can see, by comparing the similar triangles in Figure 7.13, that

$$\frac{X - x_1}{Y - y_1} = \frac{x_2 - x_1}{y_2 - y_1} \quad \text{or} \quad X = \frac{x_2 - x_1}{y_2 - y_1} \times (Y - y_1) + x_1$$

This translates into the formula given in H4.

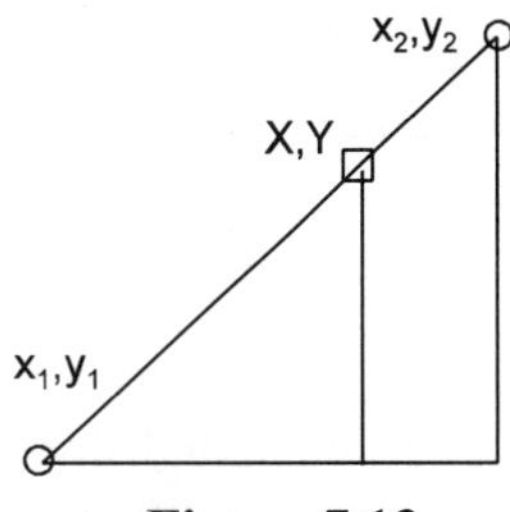

Figure 7.13

(d) To obtain the straight lines in the chart, enter these values and formulas:
A21: 0
B21: =D4
A22: =H4
B22: =D4
A23: =H4
B23: 0

You may use the Copy with Paste Special method used in Exercise 10 of Chapter 6 to make a new data series. Alternatively, right click on the chart and select *Source data*, on the *Series* tab enter Sheet7!$A$21:$A$23 for the *x*-values of a new series and Sheet7!$B$21:$B$23 for the *y*-values. These are most conveniently entered by dragging the mouse over the appropriate range. If you get curved rather than straight lines joining the three data points, right click on the line, select *Chart Type* from the popup menu, and change the type to the straight line option.

(e) Test your work by entering different values in D4. Does the value in H4 seem to be correct when you observe the chart and when you examine the raw data?

Exercise 10: Difference Formulas and Tangents

In this exercise we learn how to compute approximations to the first and second derivatives from tabulated data using the difference formulas shown below.

Order	*Forward*	*Backward*	*Central*
First	$\frac{dy}{dx}=\frac{y_1-y_0}{h}$	$\frac{dy}{dx}=\frac{y_0-y_{-1}}{h}$	$\frac{dy}{dx}=\frac{y_1-y_{-1}}{2h}$
Second	$\frac{d^2y}{dx^2}=\frac{y_2-2y_1+y_0}{h^2}$	$\frac{d^2y}{dx^2}=\frac{y_0-2y_{-1}+y_{-2}}{h^2}$	$\frac{d^2y}{dx^2}=\frac{y_1-2y_0+y_{-1}}{h^2}$

The chart in Figure 7.14 plots the data in A4:B13. If we wish to find the slope of the first point (the one nearest the origin) we could use the *y*-values for the point itself and the next point along the line using the *forward* difference formula. For the last point, we could use the *y*-values for the point itself and the previous point using the *backward* difference formula. Either of these formulas could be used for the intermediate points. However, the *central* difference formula, which uses a point before and a point after the point of interest, is more accurate.

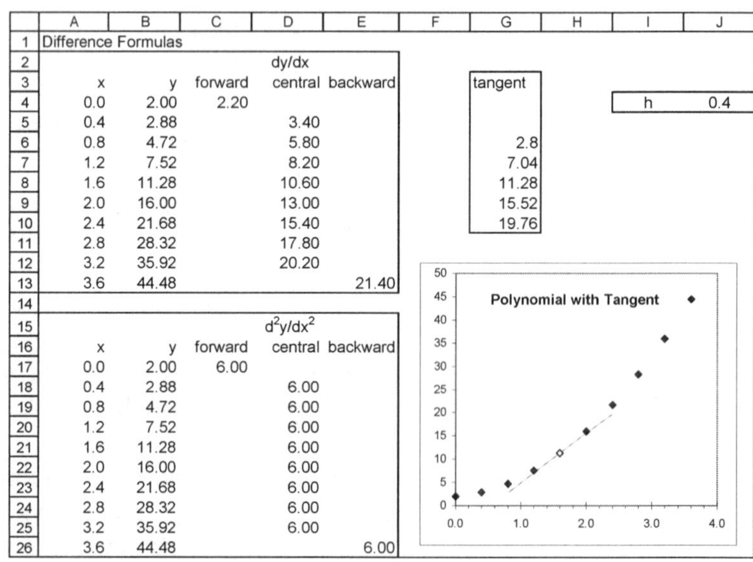

	A	B	C	D	E	F	G	H	I	J
1	Difference Formulas									
2				dy/dx						
3	x	y	forward	central	backward		tangent			
4	0.0	2.00	2.20						h	0.4
5	0.4	2.88		3.40						
6	0.8	4.72		5.80			2.8			
7	1.2	7.52		8.20			7.04			
8	1.6	11.28		10.60			11.28			
9	2.0	16.00		13.00			15.52			
10	2.4	21.68		15.40			19.76			
11	2.8	28.32		17.80						
12	3.2	35.92		20.20						
13	3.6	44.48			21.40					
14										
15				d^2y/dx^2						
16	x	y	forward	central	backward					
17	0.0	2.00	6.00							
18	0.4	2.88		6.00						
19	0.8	4.72		6.00						
20	1.2	7.52		6.00						
21	1.6	11.28		6.00						
22	2.0	16.00		6.00						
23	2.4	21.68		6.00						
24	2.8	28.32		6.00						
25	3.2	35.92		6.00						
26	3.6	44.48			6.00					

Figure 7.14

(a) Begin the worksheet on Sheet8 on CHAP7.XLS by entering the text shown in Figure 7.14. Enter the values shown in A4:B13.

(b) It will be convenient to have a cell named *h*, so enter text and value in I4:J4 and make J4 the named cell.

(c) The forward formula is implemented in C4 with =(B5 – B4)/h. Likewise, for the backward formula in E13 use =(B13 – B12)/h. In D5 the central formula is entered as =(B6 – B4)/(2*h). Be careful to remember the parentheses in the division. Copy this down to D12.

(d) The values in A17:B26 are obtained by entering =A4 in A17 and copying it across one column and down nine rows. It is left to the reader to code the formulas in C17:E26.

The constancy of the second derivative suggests the data fits a quadratic equation. Find the equation of best fit. Do the parameters of the fit give the same derivatives as our formulas?

A tangent has been drawn to the open-circled data point ($x = 1.6$) using the data in G8:G11. Let the point whose tangent we require be x_0, y_0. For a tangent we require a straight line passing through x_0, y_0 and having a slope equal to $(dy/dx)_0$. The value for y_0 in G8 comes from B8. The other points are computed from the formula $f(x_n) = f(x_0) + nh(dy/dx)_0$ where n is the number of points we have moved away from the central x_0. We have computed dy/dx in column D.

(e) The formula in G8 is =B8. In G6 we have =$B$8 – 2*h*$D$8 and in G10 =$B$8+2*h*$D$8.

(f) The data is added to the existing chart using the methods explored in Exercise 10 or Problem 2 of Chapter 6.

Problems

1. A spring of length L_0 is fixed at one end. If a force F is applied to the other end the spring will extend to length L. Hooke's law tells us that the relationship is $L = L_0 + eF$, where e is the spring's modulus of elasticity. When the spring is fixed vertically and the force is applied by attaching a body of mass m, the relationship becomes $L = L_0 + egm$, where g is the acceleration due to gravity = 9.8 m/s^2. Note that in a plot of L against m, the slope will be eg. The table below shows the results of such an experiment.

Mass (kilograms)	0.5	1	1.5	2	2.5	3
Length of spring (metres)	0.25	0.32	0.4	0.48	0.55	0.6

   Find the modulus of elasticity e and the unstretched length L_0 using:

   (a) the SLOPE and INTERCEPT functions,
   (b) an XY graph with the trendline equation, and
   (c) the LINEST array function.

   From your results in (b) or (c), comment on how well the data fits a straight line.

2. This example deals with chemical kinetics. In an experiment to determine the activation energy ΔE of a reaction, the rate constant k of the reaction was measured at various temperatures T. The variables are related by $k = A \exp(-\Delta E/RT)$, where A is an unknown constant and R, the gas constant, has the value 8.314 J·K^{-1}·mol^{-1}. By taking logarithms on both sides, we may write the relationship as $\ln(k) = \ln(A) - \Delta E/RT$. Note that the y-values will be a series of $\ln(k)$ values and the x-values will be a series of $1/T$ values.

   The table below shows the experimental results. Remember that temperature values must be converted from Celsius to Kelvin. Find the value of ΔE using the linear relationship together with:

   (a) the SLOPE and INTERCEPT functions,
   (b) an XY graph with the trendline equation, and
   (c) the LINEST array function.

Temperature (t °C)	0	10	20	30	40	50	60
Rate constant (k)	2.46E−05	1.08E−04	4.75E−04	1.63E−03	5.76E−03	1.85E−02	5.48E−02

From your results in (b) or (c), comment on how well the data fits a straight line.

3. Find the quadratic equation that best fits the data below. Make an XY plot and insert a trendline equation. You should select the Polynomial model and ensure that the value in the Order box is set to 2.

x	−2.5	−1.6	3.2	4.1
y	9.5	4.5	37.5	55

4. The data in Problem 3 fits the equation $y = ax^2 + bx + c$. Use LINEST in the way shown in Exercise 5 to find the parameters a, b and c. Add a row to your worksheet to compute the slope of the function at each point. Construct a tangent to the curve at point $x = 3.2$.

5.* In biology, the concept of isometry (constant shape) predicts that the relationship between some morphological or physiological variable, Y, and some basic size variable, X, will have the form $Y = aX^b$, where b is the *scaling exponent*. An experimenter has collected the data shown below. The basic variable is the length of the fish. Two morphological variables – length and surface area of the pectoral fin – have been measured for each fish. It is expected that the scaling constant for the length of the fin will be 1 and for the area of the fin will be 2. Make plots of log(Y) against log(X) with trendlines to test the two hypotheses. Can you suggest a method of finding the b values which does not involve computing the logarithmic values?

Length of fish (cm)	**Length of fin (cm)**	**Area of fin (cm^2)**
33.8	4.98	6.65
23.1	3.45	2.73
13.4	1.91	1.51
11.3	1.78	0.87
4.85	0.68	0.192
3.92	0.55	0.126

8
User-defined Functions

Concepts

Microsoft Excel includes a powerful programming language called Visual Basic for Applications (VBA) which enables you to write modules which may be subroutines or functions. A subroutine performs a process such as displaying a dialog box in which the user enters data. A function returns a value to a cell (or a range) in the same way as a built-in worksheet function. We shall explore only function coding. If you have experience with any programming language you will be familiar with many of the topics covered in this chapter. If you are not yet a programmer, VBA is a great way to begin. The emphasis in this chapter is on how to write functions so we will use simple examples. Later chapters make use of this skill to code more useful functions.

Visual Basic for Applications is a very broad topic and there are many books devoted to it and it alone. So you will appreciate that one chapter in this book cannot do more than give you a glimpse of its use.

Some books refer to user-defined functions as *custom functions.*

Why and when do we use *user-defined* functions? Just as it is more convenient to use =SUM(A1:A20) rather than =A1+ A2+...+A20, a user-defined function may be more convenient when we repeatedly need to perform a certain type of calculation for which Microsoft Excel has no built-in function. Once a user-defined function has been correctly coded it may be used in the same way as a built-in worksheet function.

Before you write a user-defined function, make sure that it is not already provided by Excel. The built-in functions are more efficient than user-defined functions. After you have written a function you must test it thoroughly with a wide range of input values.

Security Alert

The ability to write one's own macros is one of the outstanding features of the Microsoft Office products. However, the same features that empower the user are also available to the misguided individuals who write computer viruses. For this reason you should always be wary of accepting files from strangers. A good virus scanner is essential.

Starting with Office 2000, Microsoft has given the user additional control over what happens when a file containing a macro is opened. The options are: (1) enable macros only when they have a digital signature that you trust, (2) have the application ask if macros are to be enabled or not, and (3) trust all macros. The third option is not recommended and we do not have time to learn about digital signatures. So we shall use the middle ground.

Use the command Tool|Options and open the Security tab on the dialog box. In the lower right corner, click on the Macro Security button. Check the centre option button (Figure 8.1) then click OK.

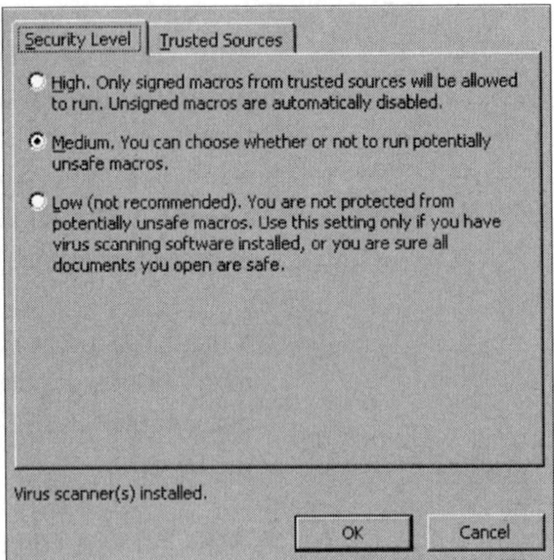

Figure 8.1

With this setting in effect, whenever you open an Excel file that contains a macro the dialog box shown in Figure 8.2 will be displayed. Clicking Enable Macros will allow you to use the functions that we are developing in this chapter.

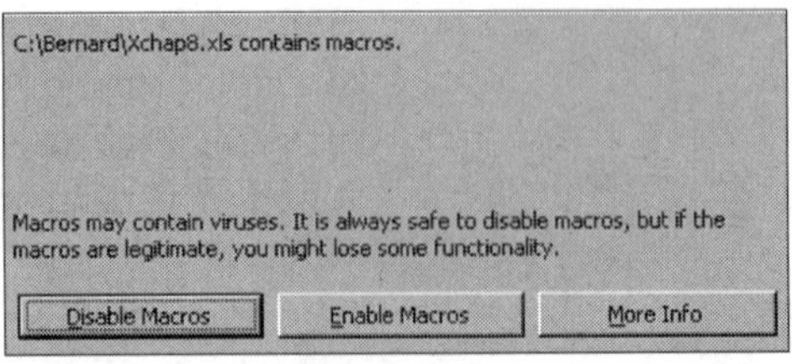

Figure 8.2

Exercise 1: The Visual Basic Editor

We will begin by briefly exploring the Visual Basic Editor (VBE) window.

(a) The VBE is reached from Excel with the command Tools|Macro|Visual Basic Editor or the shortcut Alt+F11. Figure 8.3 shows a screen capture of this window.

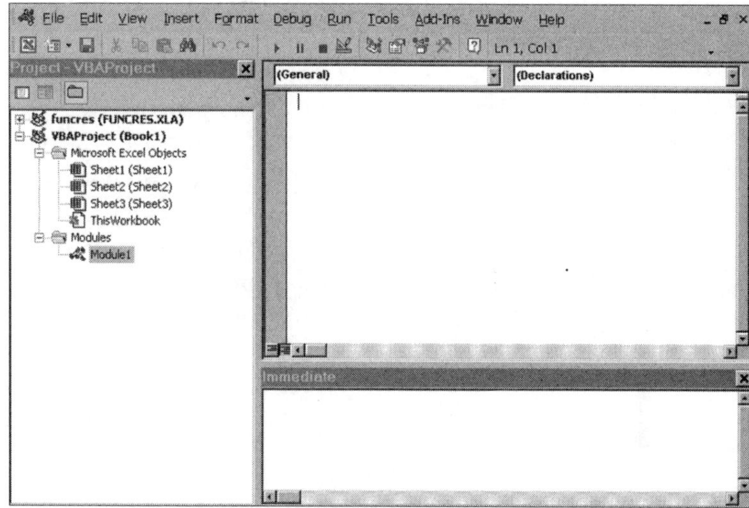

Figure 8.3

(b) As with most applications, there is a menu bar and a toolbar at the top of the window. To the right is the Project window. Yours will not be exactly the same. The figure shows two projects: one for the open Book1 worksheet, and one for the Analysis ToolPak. You will have at least the first item but its 'tree' will not display a macro item at the end – we will insert one soon.

(c) In the lower right you may see the Immediate window. If not, use the command View|Immediate Window to display it.

(d) Above the Immediate window in Figure 8.3 is the Module window but your screen currently displays a blank area. To add a module, begin by ensuring that the title *VBAProject (Book1)* is selected in the Project window. Now use the command Insert|Module from the menu bar.

(e) Return to the Excel window using one of these methods: (i) click the Microsoft Excel icon on the Windows taskbar, (ii) click the Microsoft Excel icon on the VBE toolbar, or (iii) use Alt+F11. Save the workbook as CHAP8.XLS.

Syntax for a Function

To successfully code a function you need two skills. The first is the ability to compose, in English and mathematical symbols, the set of rules that will yield the desired result. This is called the *algorithm*. The second is the ability to translate the algorithm into the Visual Basic language. Like all languages, both natural and computer, Visual Basic has a set of rules known as the language *syntax*. Figure 8.4 outlines the syntax for a user-defined function.

The optional items [*Private* | *Public*] have been omitted for simplicity.

Function name [(arglist)][**As** type]
  [statements]
  [name = expression]
  [Exit Function]
  [statements]
  [name = expression]
End Function

name	The name you wish to give to the function.
arglist	List of arguments passed to the function. Arguments are separated from each other by commas.
type	The data type of the value returned by the function.
statement	A valid Visual Basic statement.
expression	An expression to set the value to be returned by the function.

Items shown within square brackets [...] are optional.
Words in **bold** must be typed as shown.
Each statement must begin on a new line. If a statement is too long for one line, type a space followed by an underscore character and complete the statement on the next line. Do not split a word using this method.

Figure 8.4

If you edit an existing user-defined function, which is already referenced in a worksheet, that worksheet must be recalculated (press F9*) for Excel to calculate the function's new value.*

Exercise 2: A Simple Function

In this exercise we write a user-defined function to calculate the area of a triangle given the length of two sides and the included angle: Area = ½*ab* sin(θ). To test the function, a simple worksheet formula is also used to compute the area.

(a) Open CHAP8.XLS and on Sheet1 type the entries shown in A1:E3 and A4:C6 of Figure 8.5. The formula in D4 is =0.5 * A4 * B4 * SIN(RADIANS (C4)) and computes the area so that we may test our function. Copy this down to row 6. Leave E4:E6 empty for now.

	A	B	C	D	E
1	Test function to compute area of triangle				
2					
3	SideA	SideB	Angle	Formula	Function
4	1	2	90	1	1
5	2	2	45	1.414214	1.414214
6	2	2	60	1.732051	1.732051

Figure 8.5

(b) Use Alt+F11 to open the VBE window. Click on *Module1* in the Project window. The window title should read *CHAP8.XLS - [Module1 (Code)]*. One of the commonest errors for VBA beginners is entering the code in the wrong place. For our purposes, the only correct place is on a module.

```
1   ' Computes area of a triangle given two sides and the included angle
2   Function Triarea(side1, side2, Theta)
3      Alpha = WorksheetFunction.Radians(Theta)    ' Degrees to Radians
4      Triarea = 0.5 * side1 * side2 * VBA.Sin(Alpha) ' Computes area
5   End Function
```

Do not type the line numbers

Figure 8.6

(c) Having established that you are in the right place, enter the code shown in Figure 8.6 *without the line numbers* which are included here to aid later discussion. Press Enter to start each new line and Tab to indent a line. The statements in the function are explained below.

(d) The syntax of a module can be checked with the command Debug|Compile VBA Project to check for errors. You may wish to make a deliberate error to see how this feature works. Change *Sin* in line 4 to *Sine*. Now use Debug|Compile VBA

Project. The word *Sine* will be highlighted and a dialog box will inform you that this word is incorrect. Correct it before proceeding.

(e) Return to the worksheet and in E4 enter the formula =TRIAREA(A4, B4, C4). If you prefer, you may use the Function Wizard to enter this. Your function will be found in the User-defined category. Copy the formula down to row 6.

(f) The values in the D and E columns should agree. If they differ, return to the module sheet and correct the function. Remember to press F9 to recalculate the worksheet after editing a function. Save the workbook as CHAP8.XLS.

Although this is a rather simple function, it demonstrates some important Visual Basic features. We now examine each line of the TRIAREA function.

1 This is a comment line used to document the function. Visual Basic ignores any text following a single quote and displays the text in green.

2 This starts the function with the keyword Function. Keywords are displayed in blue. We have named the function *Triarea* and have declared that three arguments are used by the function. You may choose any name for your function but take care not to use the name of an existing worksheet or function. The arguments give names to the values received from the worksheet. In the worksheet, the function is invoked, or *called*, using the formula =TRIAREA(A4,B4,C4). The calling formula passes the values of its arguments to the function by position not by name. Thus the value in cell A4 is passed to the variable Side1 in the function.

3 This is an *assignment* statement; the variable *alpha* is given a value. We wish to enter the angle in degrees in the worksheet but, since we are about to use a trigonometric function, we must find the equivalent angle in radians. We could have used the statement: Alpha = Theta * 3.1416/180.0 but chose to use the RADIANS function. Since this is a worksheet function, not a VBA function, it is necessary to precede it with the keyword *WorksheetFunction* followed by a period. As soon as you have typed the period following WorksheetFunction (which may be typed without capitalization), the Editor opens a popup menu‡ listing all the available worksheet functions. You may either

‡ Should this not work, on the VB Editor menu use Tools|Options and on the *Editor* tab check the *Auto List Members* item.

type the name of the function or click a name in the list.

The complete syntax for referencing a worksheet function is *Application.WorksheetFunction.FunctionName* but the first word may be omitted as it is in our function. To maintain compatibility with earlier versions you may also use the syntax *Application.FunctionName.*

We have added documentation to this line with the comment starting with the single quote. Note that a comment may be either a line on its own or appended to the end of a statement line.

4 In this assignment statement, the function is given a value to be returned to the worksheet. Every function must contain at least one statement which assigns a value to the function.

The trigonometric *Sin* function is available in Visual Basic. We have used *VBA.Sin* in the example in order to call up the list of VBA functions. However, the statement *Triarea = 0.5* side1*side2*Sin(Alpha)* would also have been syntaxically correct.

5 This line ends the function. Every function must terminate with *End Function.*

Naming Functions and Variables

Try to use short but meaningful names for variables, functions and arguments. These three simple rules must be followed:

(i) The first character must be a letter. Visual Basic ignores uppercase and lowercase. If you use the name *term* in one place and *Term* elsewhere, Visual Basic will change the name to match the last used form.

(ii) A name may not contain a space, a period (.), exclamation point (!), @, $ or #.

(iii) A name may not be a VBA restricted keyword. Generally, VBA displays keywords in blue. A full list of these words may be found by searching in Help for 'restricted keywords'. However, it is not necessary to know them all because, if you try to use one, VBA highlights the word and displays an error message. Generally this will read 'Identifier expected' but certain keywords generate other messages.

Worksheet and VBA Functions

The mathematical functions available within VBA are shown in Figure 8.7. Details of other functions may be found by searching for *String functions* or *Date functions* in the VBA Help.

When copying formulas from VBA Help, remember to code statements in the form Arcsin = Atn(...) rather than Arcsin(X) = Atn(...).

Abs(x) The absolute value of *x*.

Atn(x) Inverse tangent of *x*. Other inverse functions may be computed using trigonometric identities such as: Arcsin(X) = Atn(X / Sqr(-X * X + 1)). For more information search Visual Basic Help for *Derived math functions.*

Cos(x) The cosine of *x*, where *x* is expressed in radians.

Exp(x) The value e^x.

Fix(x) Returns the integer portion of *x*. If *x* is negative Fix returns the first negative integer greater than or equal to *x*; for example, Fix(−7.3) returns −7. See also Int.

Int(x) Returns the integer portion of *x*. If *x* is negative Int returns the first negative integer less than or equal to *x*; for example, Int(−7.3) returns −8. See also Fix.

Log(x) The value of the natural logarithm of *x*. Note how this differs from the worksheet function with the same name which, without a second argument, returns the logarithm to base 10. In VBA, the logarithm of *x* to base *n* may be found using the statement y = Log(x)/Log(n).

Mod In Visual Basic this is an operator not a function but it is similar to the worksheet MOD function. It is used in the form number Mod divisor and returns the remainder of *number* divided by *divisor* after rounding floating-point values to integers. The worksheet function and the VBA operator return different values when the number and divisor have opposite signs – see Help for details.

Rnd(x) Returns a random number between 0 and 1.

Sgn(x) Returns −1, 0 or 1 depending upon whether *x* has a negative, zero or positive value.

Sin(x) The sine of *x*, where *x* is expressed in radians.

Sqr(x) Square root of *x*.

Tan(x) The tangent of *x*.

Figure 8.7

You cannot use a worksheet function when VBA provides the equivalent function even when the name is not the same. So none of the worksheet trigonometric functions *SIN*, *COS* or *TAN* may be used but *ASIN* and *ACOS* are permitted. The worksheet function *SQRT* cannot be used since VBA includes the equivalent *SQR* function. You may, however, use the worksheet function *MOD* because *Mod* in VBA is an operator not a function.

Use the Help facility in the VB editor to see a list of which worksheet functions are available for use in VBA code. On the *Answer Wizard* tab, type *List of worksheet functions* in the *Search* box. Alternatively, you may use the Auto List feature mentioned in Exercise 2. Typing WorksheetFunction. on a module sheet brings up a list of worksheet functions while typing VBA. brings up a list of VBA functions.

If you try to code a user-defined function that references an unavailable function (e.g. WorksheetFunction.Sin(Alpha)), the worksheet cell in which your user-defined function is called will display #VALUE!. The #NAME? error value is returned if your function uses a function from another workbook or from the Analysis ToolPak unless you follow the instructions in *Using Functions from Other Workbooks* at the end of this chapter.

Exercise 3: When Things Go Wrong

Recalling the adage 'To err is human, but to really mess up you need a computer', we will make an error in a module and lock our worksheet. Do not worry, we can fix the problem. It is very likely that you will accidentally make such an error, so it is good to know what is needed to correct it.

(a) Open the VBA Editor and change line 4 by replacing the equal sign by a minus sign. Do not press [Enter↵] and do not use Debug|Compile. Just return to the worksheet.

(b) Press [F9] to recalculate the worksheet. Microsoft Excel returns you to the VBA Editor window and displays an error dialog box. Click the *OK* button. Note that the function header is highlighted in yellow. Correct the error by replacing the minus sign by an equal sign. The yellow highlighting does not disappear.

(c) Return to the worksheet. Try to do something like changing the active cell. Nothing works; the worksheet (indeed the whole workbook) is locked.

(d) Return to the VBA Editor and use the command Run|Reset to remove the highlighting. Now when you go to the worksheet and press [F9] all is well again.

In short: an error in a module can cause a worksheet to lock. The remedy is to use the command Run|Reset to reset the module.

This exercise should help you with syntax error. Another type of problem is the logical error. This occurs when you have coded your function correctly as far Visual Basic is concerned (the syntax is correct) but the wrong answer results from an error in the algorithm. Some techniques for solving this type of problem are explored in Exercise 12.

Programming Structures

The normal flow in any computer program (and our function is a small computer program) is from line to line. This is called a *sequential structure.* In Exercise 2 (Figure 8.6) line 3 is executed, followed by line 4, then line 5. Anything that changes this flow is called a *control structure*. In the next exercise we look at a *branching* or *decision structure*. This structure gives the program two or more alternative paths to follow depending on the value of a variable. The other control structure is the *repetition* or *looping structure* which we explore in later exercises. The code within a loop is executed one or more times.

Exercise 4: The IF Structure

The function $\sin(x)/x$ presents a minor problem in that we cannot compute its value by simple division when $x = 0$. However, it is known that $\sin(x)/x = 1$ when $x = 0$. So we need one method for the calculation when $x = 0$ and another method for all other cases. On a worksheet we would use =IF(A1=0, 1, SIN(A1)/A1). The same type of construction is used in a macro using the IF...ELSE structure whose syntax is shown in Figure 8.8. The items enclosed in square brackets are optional. For details on *conditions* refer back to Chapter 5. There is a slightly simpler syntax which may be used when the IF statement will fit on one line.

(a) On Sheet2 of CHAP8.XLS enter the text in A1:B3 and the values in A4:A14 as shown in Figure 8.9.

(b) In B4 enter the formula =FSIN(A4). Because we have yet to write a function named *Fsin*, the error value #NAME? is displayed in B4. This problem will be solved shortly. Copy B4 down to B14.

Syntax for IF...ELSE

If condition **Then**
    [statements]
[**ElseIf** condition-n **Then**
    [elseifstatements]] . . .
[**Else**
    [elsestatements]]
End If

condition	Expression that is True or False.
statements	One or more statements executed if condition is True.
condition-n	Numeric or string expression that evaluates True or False.
elseifstatements	One or more statements executed if associated condition-n is True.
elsestatements	One or more statements executed if no previous condition-n expressions are True.

Figure 8.8

	A	B
1	The Sin(x)/x function	
2		
3	x	Sin(x)/x
4	1	0.841471
5	0.8	0.896695
6	0.6	0.941071
7	0.4	0.973546
8	0.2	0.993347
9	0	1
10	-0.2	0.993347
11	-0.4	0.973546
12	-0.6	0.941071
13	-0.8	0.896695
14	-1	0.841471

Figure 8.9

```
Function fsin(x)
    If x = 0 Then
        fsin = 1
    Else
        fsin = Sin(x) / x
    End If
```

Figure 8.10

(c) Use Alt+F11 to open the VBA Editor window. Select CHAP8.XLS in the Project window and use Insert|Module. The 'Tree' for the current project will now show Module1 and Module2. We discuss below why we have done this.

(d) On the Module2 code sheet, type the function shown in Figure 8.10 without the line numbers. Return to the worksheet and press F9 to recalculate the worksheet.

To learn about the shorter syntax for IF, go to the module you have just coded, move the cursor to the word *If* and press F1 to activate the context-sensitive Help for Visual Basic for Applications.

We made a second module for the user-defined function of this exercise. This was not essential, we could have added it to Module1. However, there is a problem with having more than one function on a single module. If any one of the functions contains an error then all functions on that module return error values on the worksheet. This can be confusing, especially for beginners. There are other considerations that help you decide whether to put more than one function on a single module but these relate to the use of *Public* or *Private* in the Function header – a topic we will not be exploring.

Exercise 5: Boolean Operators

In this exercise we will use the short and the long form of the IF statement, and show the use of the AND and OR Boolean operators.

We will construct a function which reports what type of triangle is formed when given the length of the three sides. Before coding this we need to think more about the algorithm. We might make a list of the things we know about triangles and their sides:

(i) One side is always shorter than the sum of the other two.

(ii) In an equilateral triangle all the sides are equal.

(iii) In an isosceles triangle two sides are equal.

(iv) Pythagoras' theorem is true with a right angle triangle.

How do we know which is the hypotenuse if we are given just three values? How many pairs of sides must be compared to establish we have an isosceles triangle? These questions are readily answered if the values for the sides are in descending order. The reader may wish to complete the algorithm before proceeding.

(a) On Sheet3 of CHAP8.XLS construct the worksheet shown in Figure 8.11. The cell D4 contains the formula =Tritype(A4, B4,

C4). It will return the value #NAME? until we have coded the function.

	A	B	C	D
1	What type of triangle?			
2				
3	Side A	Side B	Side C	Type
4	2	1	2	Isosceles

Figure 8.11

(b) Move to the VBA window, add a third module to the CHAP8.XLS project and code the function shown in Figure 8.12.

(c) Return to the worksheet and press F9. Experiment with other values for the three sides to test the function.

```
Function Tritype(a, b, c)
    If b > a Then temp = a: a = b: b = temp
    If c > a Then temp = a: a = c: c = temp
    If c > b Then temp = b: b = c: c = temp
    If a > b + c Then
        Tritype = "None"
    ElseIf a * a = b * b + c * c Then
        Tritype = "Right"
    ElseIf (a = b) And (b = c) Then
        Tritype = "Equilateral"
    ElseIf (a = b) Or (b = c) Then
        Tritype = "Isosceles"
    Else
        Tritype = "Scalene"
    End If
End Function
```

Figure 8.12

The function begins with three IF statements to sort the sides into descending order to make it easier to test for the type of triangle. Because each IF statement is written on a single line, no ENDIF is required. Note how each IF structure contains three assignment statements separated by colons. The longer form of the IF statement is used to test the triangle. Two examples of the use of Boolean operators are shown. Note that sorting the values of *a*, *b* and *c* in the function has no effect on the values in the cells A1:A3. We say that variables within the function have a *local scope*.

Exercise 6: The SELECT Structure

Visual Basic for Applications provides another branching structure called the SELECT CASE structure. Its syntax is shown in Figure 8.13.

The keyword *To* specifies a range of values and has the form *smaller-value To larger-value*. The *Is* keyword is used with comparison operators (e.g. >, < , >=, <=) to specify a range of values.

When *testexpression* matches one of the expressions, the statements following that Case clause are executed up to the next Case clause, or, for the last clause, up to the *End Select*. Control then passes to the statement following *End Select*. When *testexpression* matches an expression in more than one Case clause, only the statements for the first match are executed.

Syntax of SELECT CASE

```
Select Case testexpression
[Case expressionlist-n
    [statements-n]] . . .
[Case Else
    [elsestatements]]
End Select
```

testexpression	Any numeric or string expression.
expressionlist-n	A list of one or more of expression types separated by commas. Valid expression types are: expression, expression To expression, Is comparisonoperator expression.
statements-n	One or more statements executed if testexpression matches any part of expressionlist-n.
elsestatements	One or more statements executed if testexpression does not match any of the Case clauses.

Figure 8.13

The statements following *Case Else* are executed if no match is found in any of the other Case selections. It is advisable always to use a *Case Else* statement to handle unexpected *testexpression* values.

The IF statement in the *Fsin* function could be replaced by a SELECT CASE statement in the form:

```
Select Case x
    Case 0      fsin = 1
    Case Else   fsin = sin(x)/x
End Select.
```

The longer IF statement in *Tritype* could not be coded as a SELECT CASE since that structure allows only one test expression.

The number of real roots of the quadratic $ax^2 + bx + c = 0$ is determined by the value of the discriminant $d = b^2 - 4ac$. When d is positive there are two real roots, one when it is zero and none when it is negative. In this exercise we write a function to return a value indicating the number of real roots for a quadratic equation.

(a) Open the VB Editor, insert another module and enter the function shown in Figure 8.14.

Reminder: When you return to the worksheet after editing a user-defined function, the worksheet will not recognize that the UDF has been altered until it performs a recalculation. This will happen whenever the value in a cell is changed but the F9 key may be used to force a recalculation.

```
Function RootCount(a, b, c)
    d = (b * b) - (4 * a * c)
    Select Case d
        Case 0
            RootCount = 1
        Case Is > 0
            RootCount = 2
        Case Else
            RootCount = 0
    End Select
End Function
```

Figure 8.14

	A	B	C	D
1	Number of real roots of a quadratic equation			
2				
3	a	b	c	Roots
4	1	0	-9	2
5	1	-5	6	2
6	1	2	8	0
7	1	-8	16	1

Figure 8.15

(b) On Sheet4 of CHAP8.XLS enter the text and values in A1:D3 of Figure 8.15 and the values in A4:C7. In D4 enter =RootCount(A4,B4,C4) and copy it down to row 7. The values in column D should agree with those in Figure 8.15.

(c) Constant values may be used as arguments. To demonstrate this, move to an empty cell such as A10 and enter =RootCount(1,-5,6). This should return the value 2.

(d) Move to an empty cell such as A11 and enter =RootCount(A3,B3,C3). This will return the error value #VALUE!. There are a number of conditions that result in this value. We have passed non-numeric arguments (the cells A3, B3, C3 contain text) to a function expecting numeric values. This error can also be caused by a syntax error in the function.

Exercise 7: The FOR...NEXT Structure

In a looping structure a block of statements is executed repeatedly. When the repetition is to occur a specified number of times, the FOR...NEXT structure is used. The syntax for this structure is given in Figure 8.16.

Microsoft Excel provides the function FACT to compute *n* factorial. For example, =FACT(4) returns the value of 4! = 1 × 2 × 3 × 4. In its absence, we could have coded such a function using a FOR loop:

```
Function Fact(n)
    Fact = 1
    For j = 1 to n
        Fact = Fact * j
    Next j
End Function.
```

When the loop is first entered *j* is given a value of 1. The statement Fact = Fact * j may be confusing at first. Remember that computer code statements are instructions, not mathematical equations. This statement may be read as: the new value for Fact is its old value multiplied by 1. The Next statement increments (increases) j by 1 and sends the program back to the loop. Another new value of Fact is computed as Fact *2. These steps are repeated until j is greater than the argument n at which point the program continues to the line following the Next statement. The reader should see that the function would be slightly more efficient if the third line read For j = 2 to n.

There is also in VBA a *FOR EACH* structure which is explored in Problem 5 at the end of the chapter.

Syntax of FOR...NEXT

For counter = first **To** last [**Step** step]
    [statements]
    [**Exit For**]
    [statements]
Next [counter]

counter	A numeric variable used as a loop counter.
first	The initial value of counter.
last	The final value of counter.
step	The amount by which counter is changed each time through the loop.
statements	One or more statements that are executed the specified number of times.

The step argument can be either positive or negative. If step is not specified, it defaults to one.

After each execution of the statements in the loop, step is added to counter. Then it is compared to last. When step is positive the loop continues while counter <= end. When step is negative looping continues while counter >= end.

The optional Exit For statement, which is generally part of an IF statement, provides an alternate exit from the loop.

Note: If the value of *counter* is altered by a statement anywhere between the *For* and the *Next* statements, you run the risk of unpredictable results.

Figure 8.16

For this exercise we will write a function to compute the sum of the squares of the first n integers. Our function will find the value of $1^2 + 2^2 + 3^2 + \ldots + n^2$ or $\sum_1^n k^2$.

(a) Insert Module5 into the CHAP8.XLS project. Code the function shown in Figure 8.17.

(b) On Sheet5 construct a table similar to that in Figure 8.18 to test your function.

```
Function SumOfSquares(n)
  SumOfSquares = 0
  For j = 1 To n
    SumOfSquares = SumOfSquares + j ^ 2
  Next j
```

Figure 8.17

	A	B
1	n	SumOfSquares
2	1	1
3	2	5
4	3	14
5	4	30
6	5	55

Figure 8.18

The first statement in our function is not really needed because Visual Basic initializes all numeric variables to zero. However, it does no harm to perform the initialization explicitly and can make it easier for another user to understand the code.

Exercise 8: The DO...LOOP Structures

Whereas the FOR...NEXT structure is used for a specific number of iterations through the loop, the DO...LOOP structures are used when the number of iterations is not initially known but depends on one or more variables whose values are changed by the iterations.

The first syntax of the DO structure is shown in Figure 8.19.

Syntax 1 for the DO statements

Do {**While** | **Until**} condition
  [statements]
  [**Exit Do**]
  [statements]
Loop

Figure 8.19

From this syntax we see there are two variations.

1. *DO WHILE condition...LOOP*: Looping continues while the condition is true. The condition is tested when the first line of the DO statement is executed.

2. *DO UNTIL condition...LOOP*: Looping continues until the condition is true. The condition is tested when the first line of the DO statement is executed.

Note that the condition is tested at the start of the structure. If the condition is false, none of the statements in the loop are executed.

The second syntax tests the condition at the end of the structure – Figure 8.20.

Syntax 2 for the DO statements

Do
    [statements]
    [Exit Do]
    [statements]
Loop {**While** | **Until**} condition

Figure 8.20

This gives rise to two variants:

3. *DO...LOOP WHILE condition*: Looping continues while the condition is true. The condition is tested when the last line of the DO statement (the line beginning with LOOP) is executed.

4. *DO...LOOP UNTIL condition*: Looping continues until the condition is true. The condition is tested when the last line of the DO statement is executed.

Because the condition is tested at the end of the structure, the statements within the loop are executed at least once regardless of the value of the condition at the start of the DO structure.

The two examples that follow show the difference between Syntax 1 and 2. With Code A, the statements within the loop will not be executed since t has a value less than 0.01 when the condition is tested. With Code B, the loop will be entered and the statements will be executed until $t < 0.01$. This is true regardless of the value assigned to t in the first statement.

Code A
```
t = 0
n = 1
sum = 0
DO UNTIL t < 0.01
     t = 1 / n
     sum = sum + 1 / t
     n = n + 1
LOOP
```

Code B
```
t = 0
n = 1
sum = 0
DO
     t = 1 / n
     sum = sum + 1 / t
     n = n + 1
LOOP UNTIL t < 0.01
```

Typically, the *condition* in the UNTIL or WHILE phrase refers to one or more variables. The programmer is responsible for ensuring that the variables eventually have values such that the exit condition is satisfied. The only exception is when a conditional EXIT statement is used to terminate the loop.

Consider the code:

```
t = 0
DO UNTIL t > 0.001
     statement-1
     ...
     statement-n
LOOP
```

Unless there is an EXIT DO within the loop, at least one of the statements must alter the value of *t* in such a way that, after a finite number of iterations, *t* has a value greater than 0.001. If this is not the case the loop will execute unendingly.

If you happen to code an infinite loop by mistake, your worksheet will 'hang'. Use Ctrl+Break *to abort the function.*

Have you ever wondered how your calculator, or an application such as Microsoft Excel, 'knows' the value of such quantities as sin(0.55) or exp(0.456)? The answer of course is that they are calculated from power series. Two examples of Maclaurin series are shown below.

$$\sin(x) = \sum_{k=0}^{\infty} \frac{(-1)^k x^{2k+1}}{(2k+1)!} = x - \frac{x^3}{3!} + \frac{x^5}{5!} - \frac{x^7}{7!} \cdots$$

$$\exp(x) = \sum_{k=0}^{\infty} \frac{x^k}{k!} = 1 + x + \frac{x^2}{2!} + \frac{x^3}{3!} \cdots$$

Of course, the summations cannot be carried on to infinity. The beauty of these series is that they converge; the terms get progressively smaller and smaller. So we may use as many terms as we need for a specified degree of precision.

In this exercise we will use a DO…LOOP to code a function that computes exp(x) using the power series shown above. We note that:

$$\text{term}_k = \frac{x^k}{k!} \quad \text{and} \quad \text{term}_{k+1} = \frac{x^{k+1}}{(k+1)!}$$

$$\therefore\ \text{term}_{k+1} = \text{term}_k \times \left(\frac{x}{k}\right)$$

Thus, we can compute each successive term from the one before it.

```
 1 Function MacExp(x)
 2    Const Precision = 0.000000000001
 3    MacExp = 0
 4    Term = 1
 5    k = 0
 6    Do While Term > Precision
 7       MacExp = MacExp + Term
 8       k = k + 1
 9       Term = Term * x / k
10    Loop
11 End Function
```

Figure 8.21

	A	B
1	Maclaurin series for Exp(x)	
2		
3	x	2
4	MacExpt	7.3890560989301700
5	Excel Exp	7.3890560989306500
6	Difference	-4.760636330E-13

Figure 8.22

(a) On Module6 enter the code shown in Figure 8.21 – without the line numbers, of course. The value in line 2 is entered as 1e-12; VBA changes it to the form shown. We have set Precision as a constant. Walk through the program and follow it for three or four loops to ensure that you follow its workings.

(b) Construct a testbed on Sheet6 similar to that shown in Figure 8.22. Experiment with large and small values in B3. What do you find? Change the value of Precision to 1E−16 and try large and small numbers again.

(c) Did you remember to try negative values for *x*? Recall the advice previously given: thoroughly test every computer program you write. So what happens when *x* is negative? Can you explain why? The solution is to change line 6 to read: Do While abs(Term) > Precision. When the argument is negative the second term is negative and this is smaller than Precision so the loop is terminated after one iteration until we test the absolute value of Term.

Variables and Data Types

Unlike most computer languages, all dialects of BASIC allow the programmer to use variables without first declaring them. While this feature slightly speeds up the coding process, it has the major disadvantage that typo errors can go undetected. Suppose a function uses the variable named Angle but on one line the programmer types Angel. Visual Basic will treat these as two different variables and if the error occurs in a long piece of code it can be difficult to spot. The problem is avoided by the use of the Option Explicit statement at the start of the module to force explicit declarations of all variables of every function in the module sheet. With this in place, all variables must be declared before being used. We will use the DIM statement for this purpose.

You can arrange to have *Option Explicit* automatically added to every new module. Open the Tools|Options dialog box in VBE and on the Editor tab place a check mark beside *Require Variable Declaration.*

```
Option Explicit
Function MacExp(x)
   Dim Term, k
   Const Precision = 0.000000000001
   MacExp = 0
   Term = 1
   k = 0
   Do While Term > Precision
      MacExp = MacExp + Term
      k = k + 1
      Term = Term * x / k
   Loop
End Function
```

Figure 8.23

Figure 8.23 shows the function MacExp from Exercise 8 recoded to demonstrate this. We do not use the function name MacExp, the argument x, or the constant Precision in the dimension (Dim) statement. The Function statement declares the function name and arguments, while the Const statement declares its own variable.

Suppose that line 11 had been mistyped as term = tern * x / k. The cell using this function would show the error value #NAME? In the module sheet, the word *tern* would be highlighted and the error message 'Variable not declared' would be displayed.

You may wish to consider using this feature in all your functions or only when the code is long. Remember that after editing a module, the worksheet must be recalculated by pressing F9.

There is yet another difference from other programming languages. In languages such as FORTRAN, C, etc., it is not sufficient merely to name the variable; the programmer must also state its data type. In this example we would need to define *k* as an integer variable and *term* as a floating-point variable. We could do this by coding line 3 as Dim term As Double, k As Integer. When the data type of variables are not declared Visual Basic uses a special data type called *variant.* This is acceptable for the simple functions shown in these examples but in general one should declare data types. It should be noted that variant data types are memory hogs.

You may wish to use Help to find out more about this topic, especially the permitted range of values for Integer, Short, Long, Single and Double.

Exercise 9: A User-defined Array Function

We may require our user-defined function to return more than one value or we may wish to pass a range of cell values to a function. In this exercise we do both. We will code a function to compute the real roots of the quadratic equation $ax^2 + bx + c = 0$.

As an alternative to the Option Base 1 declaration you could use Dim Temp(1 to 3).

(a) Insert Module7 in the CHAP8.XLS project and code the function shown in Figure 8.24.

(b) We will test the function on Sheet7 as shown in Figure 8.25. Only B6:D6 have formulas. Select B6:D6 and type the formula =Quad(B5, C5, D5). This function returns an array, so complete the entry with Ctrl+Shift+Enter.

```
1 Option Base 1
2 Function Quad(a, b, c)
3       Dim Temp(3)
4       d = (b * b) - (4 * a * c)
5       Select Case d
6             Case Is < 0
7                 Temp(1) = "No real"
8                 Temp(2) = "roots"
9                 Temp(3) = ""
10            Case 0
11                Temp(1) = "One root"
12                Temp(2) = -b / (2 * a)
13                Temp(3) = ""
14            Case Else
15                Temp(1) = "Two roots"
16                Temp(2) = (-b + Sqr(d)) / (2 * a)
17                Temp(3) = (-b - Sqr(d)) / (2 * a)
18      End Select
19      Quad = Temp
20 End Function
```

Figure 8.24

	A	B	C	D
1	Quadratic Equation Solver			
2				
3		$ax^2 + bx + c = 0$		
4		a	b	c
5	Coeff	1	5	6
6	Roots	Two roots	-2	-3

Figure 8.25

(c) Try other values for *a*, *b* and *c* to test the function. Solve these equations:
(i) $x^2 + 3x + 3 = 0$ (no real roots),
(ii) $x^2 - 4x + 4 = 0$ (one real root).

The function contains some new features:

(i) VBA does not permit us to declare a user-defined function as an array so we use a temporary array to hold the three values until the end of the function.

(ii) Note how an array is defined in line 3. This line declares three variables with the name *temp*. We distinguish one from the other using an index. Thus we may refer to *temp(1)*, *temp(2)* and *temp(3)*. In the absence of the *Option Base 1* statements, these would have been *temp(0)*, *temp(1)* and *temp(2)*.

(iii) Values are entered into the elements of the array in lines 7 to 17 depending on the value of the discriminant. Some expressions contain parentheses which are not essential but aid the reading of the expressions.

(iv) In line 19, the values of the array are passed to the function to be returned to the worksheet.

The call to the function uses the formula =Quad(B4, B5, B6). If you wish to call the function using =Quad(B4:D4), the function could be recoded as shown below.

```
Function Quad(coeff)
   Dim Temp(3)
   d = (coeff(2) ^ 2) - (4 * coeff(1) * coeff(3))
   Select Case d
      Case Is < 0
         Temp(1) = "No real"
         Temp(2) = "roots"
         Temp(3) = ""
      Case 0
         Temp(1) = "One root"
         Temp(2) = -coeff(2) / (2 * coeff(1))
         Temp(3) = ""
      Case Else
         Temp(1) = "Two roots"
         Temp(2) = (-coeff(2) + Sqr(d)) / (2 * coeff(1))
         Temp(3) = (-coeff(2) - Sqr(d)) / (2 * coeff(1))
   End Select
   Quad = Temp
End Function
```

Exercise 10: Inputting an Array

In the last part of the previous exercise we saw how to pass a one-dimensional array to a function. This could be a group of contiguous cells in a row or in a column. In this exercise, we pass a two-dimensional array to a function. We pass a block of 12 cells arranged in three rows and four columns to the function which sums each of the three rows and returns the value of the three sums.

> The *FOR EACH* structure explored in Problem 5 provides another way of using ranges as arguments for a UDF.

(a) In the VB Editor, insert Module8 and code the function shown in Figure 8.26.

(b) On Sheet8, enter the values shown in Figure 8.27 excluding D7. In D7 enter =MaxRow(A3:D5). Check that the function is returning the correct value and save the workbook.

```
1 Option Base 1
2 'Function to return maximum sum of three rows
3 Function MaxRow(Data As Object)
4       Dim RowSum(3)
5       For r = 1 To 3
6           For c = 1 To 4
7               RowSum(r) = RowSum(r) + Data.Cells(r, c)
8           Next c
9    Next r
10   MaxRow = Application.Max(RowSum)
11 End Function
```

Figure 8.26

	A	B	C	D
1	Maximum Row Value			
2				
3	12	14	12	6
4	4	15	6	6
5	9	9	12	8
6				
7	Largest row sum value			44

Figure 8.27

The code *Data As Object* in the argument list of the function references range A3:D5 in the worksheet formula. In line 4 the array *RowSum* is created to hold the three values of the sum of each row. The nested FOR loops compute each sum. Note the use of *Data.Cell(r,c)* to reference the value of each cell in the range. In line 10, the worksheet MAX function is used to find the largest value in the *RowSum* array.

Exercise 11: Improving Insert Function

The functions you have written will appear on the Insert Function dialog box in the *User-defined* category but we need to make a small improvement.

(a) Make E6 on Sheet1 the active cell and delete its formula.

(b) Click on the Insert Function tool. Select the *User-defined*

Insert Function tool

category and you will see the names of all the functions in this workbook. Select the Triarea function. The lower part of the dialog box mistakenly states *Choose the Help button for help on this function and its arguments*. Click on the *Cancel* button.

(c) Use the command Tools|Macro|Macros (or the shortcut Alt+F8) to open the Macro dialog box. In the *Macro name* box enter the function name *Triarea*. Click on the *Option* button. In the *Description* box enter a short description of the functions. Click OK. The description is displayed at the bottom of the Macro dialog box. Close the box using the X in the top right corner.

(d) Use the Insert Function tool to replace the formula = Triarea(A6, B6, C6) in E6. Note that the Insert Function and the Function Arguments dialog boxes now display the more useful information about the function.

Exercise 12: Some Debugging Tricks

What do you do when a function returns the wrong value? The first thing to do is print it out and carefully work through it. The second thing is to find a sympathetic friend, she does not need to be a programmer, to whom you can explain the function. Many programmers have found that the act of vocalizing makes you think more clearly. You may find yourself saying to a very puzzled friend who has said nothing 'Oh! Now I see what's wrong. Thanks for your help.' When all else fails we need some serious debugging tools. Very often the problem is resolved by examining intermediate values.

The function *Tritype* which we developed in Exercise 5 requires that the three sides be sorted in descending order. If we make a mistake in the code for that part of the algorithm the function will fail. In this exercise we look at two methods of checking this part of the code.

(a) On Sheet3 enter the value 1 for all three sides of the triangle.

(b) Open the VB Editor and in the Project Explorer window double click on Module3 to bring it up in the Code window.

(c) Immediately following line 4 (see Figure 8.12) enter the statement: MsgBox "a = " & a & " b = " & b & " c = " & c.

This causes the function to display a Message box displaying the values of the three sides after the sorting has been done.

The items in quotation marks are displayed as text, the values of the variables *a*, *b* and *c* are displayed as numbers. The ampersands are used to concatenate (join) the textual and numeric values. Search the term *Msgbox* in the VBE Help to learn more about the Message box.

(d) Return to Sheet3 and enter the values 4, 3 and 5 in A3, B3 and C3, respectively. The function is recalculated as each value is entered and each time the function runs it displays the message box. The third time the box resembles Figure 8.28 and we see that the three values are indeed sorted correctly.

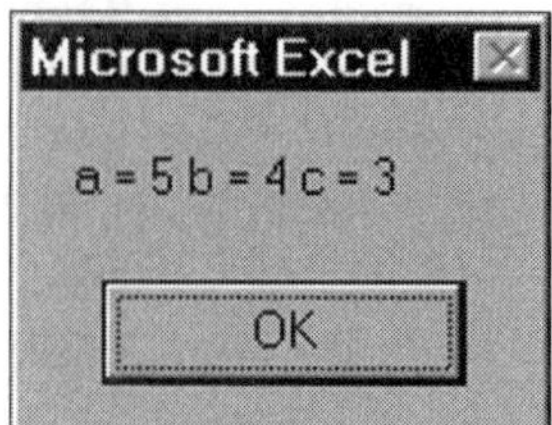

Figure 8.28

(e) In preparation for another debugging method, place a single quote in front of the newly entered code in Module3 to make it a comment. Replace each value in A3:C3 by 1.

(f) Return to Module3 and edit the new line to read:

```
Debug.Print "a = " & a & " b = " & b & " c = " & c
```

If the Immediate window is not visible use the View menu to bring it up.

(g) Return to Sheet3 and enter the values 4, 3 and 5 in A3, B3 and C3, respectively. The function is recalculated as each value is entered and each time the function runs it prints some output in the Immediate window. Open the VBE and the Immediate window should resemble that in Figure 8.29 and, again, we see that the three values are indeed sorted correctly.

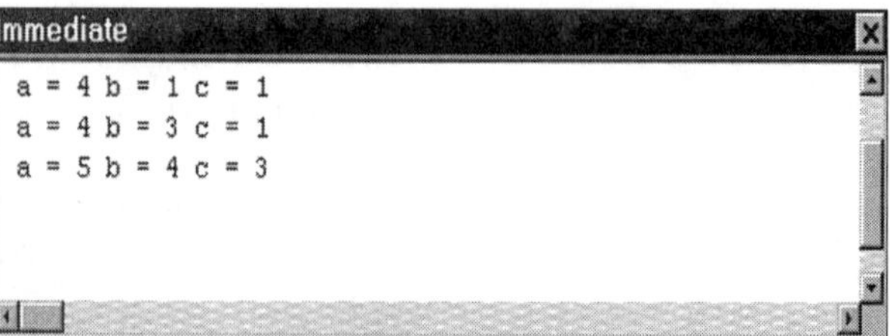

Figure 8.29

Using Functions from Other Workbooks

The user-defined functions we have created have been used in the workbook in which they were coded. Every function in a workbook is available from any sheet in it. There are a number of procedures which permit us to use a function from another workbook.

1. The least efficient method is to copy and paste the function to the new workbook.

2. The user-defined functions of an open workbook are available in other workbooks. Thus if CHAP8.XLS is open, then in a second workbook we may code, for example, =CHAP8.XLS!Quad(…). We could either type this formula or use the Function Wizard to locate the function in the User-defined category.

   XLSTART is a subfolder of the folder EXCEL in which Excel was installed, generally this will be C:\Program Files\Microsoft Office\Office10\Excel.

   If the workbook containing the modules is placed in the XSTART folder it will automatically open every time Microsoft Excel is started. To make it less obtrusive, use Window|Hide before you save the workbook. When it is opened next time it will be invisible. A hidden file can be made visible again from the Window menu.

3. If the functions of workbook CHAP8.XLS are needed in only a few other workbooks the registration method may be more appropriate. Open workbook CHAP8.XLS and the workbook (let's call it Book1) in which you wish to use its functions. In Project Explorer window of the VB Editor, click on *VBAProject (CHAP8.XLS)* and use Tool|VBAProject Properties to open the *Properties* dialog box. By default, all projects are named *VBAProject*, but we need a unique name such as *Chap8.* In the Project Explorer, click on the heading for the Book1 project and use Tools|Reference. In the resulting dialog box locate the *CHAP8* project and click the box beside it. This reference causes CHAP8.XLS to be opened automatically whenever you open Book1.

   You are not required to give the project the same name as the workbook, but it is advisable.

4. Finally, we look at the steps needed to make an add-in from CHAP8.XLS once all the functions have been coded and thoroughly tested.

(i) In the Project Explorer window, right click on *Chap8* (this assumes you renamed it in Step 3 above) and open the Properties dialog box. To prevent others from copying or modifying your code, enter a password on the *Protection* tab.

(ii) Return to the Excel window and use File|Properties. On the *Summary*, give the file a descriptive title and add a documentary remark in the *Comment* field.

(iii) Use File|Save As and in the *Type* box select *Microsoft Excel Add-in* (it is the last item). The file will be saved with the extension XLA.

(iv) You, and others to whom you give the file, will be able to use it in much the same way as you are using the Analysis ToolPak add-in.

A number of companies market Microsoft Excel add-ins. For example, one add-in contains functions for performing mass-mole chemistry calculations. You may find some shareware add-ins by searching the World Wide Web.

The procedures above allow you to use worksheet formulas that refer to functions which are coded in another workbook. Other procedures are needed if you wish to use functions from another workbook in one of your user-defined functions. To use one of the CHAP8.XLS functions within a function coded in Book1 three steps are necessary:

1. The VPAProject of CHAP8.XLS must be given a unique name such as *Chap8* – see 3 above.

2. With Book1 as the active project in the VB Editor, use Tools|Reference to reference the Chap8 project.

3. Within your code, use statements in the form *z* = *[Chap8].Macexp*, or *z* = *Macexp*. The first format is preferred for two reasons: it provides better documentation and it invokes the Auto List Member feature.

Similar procedures allow you to use the functions provided by the Analysis ToolPak within your own functions.

1. On the worksheet, use Tools|Add-ins to load *Analysis ToolPak VBA* (note the *VBA* ending).

2. With Book1 as the active project in VBE, reference *atpvbaen*.

3. Use code in the form *z = [atpvbaen].Lcm(numb1, numb2)*.

Problems

1.* Write functions with the following properties:

(a) When A1 contains a temperature value in Fahrenheit,= Kelvin(A1) returns the equivalent temperature on the Kelvin scale. Test the function using 32°F = 273.15 K and -459.67°F = 0 K.

(b) When A2 contains a number n where $n >> 1$, =Stirling(A2) returns Stirling's approximation: $\ln(n!) \approx n\ln(n) - n$. Use the worksheet FACT function to evaluate the accuracy of Stirling's approximation.

(c) When A3 and B3 contain two integer values a and b, the formula =SumRange(A3,B3) returns

$$\sum_{n=a}^{b} n \text{ when } a \le b \text{ or, } \sum_{n=b}^{a} n \text{ otherwise}$$

2.* In the Quad function of Exercise 6, line 3 defines a one-dimensional array Temp which is passed to the function Quad and returned to the worksheet. In the worksheet the values are displayed in three cells in a vertical arrangement. Thus the code Temp(3) was interpreted as one row and three columns. We could have made this explicit with Temp(1,3) and changed the references to Temp(1,1) = "No real", Temp(1,2) = "roots", etc. Use this information to write a modification of Quad (call it Quad2) which could be used as an array formula in G2:G4 to give a result similar to the figure below.

	F	G
1	Coeff	Roots
2	1	Two roots
3	-5	3
4	6	2

3. Write a function to compute the cross-product of two vectors. It should be called using =CrossProd(VecA,VecB), where VecA and VecB are each a vertical range of three cells. This is to be an array formula returning the three values in a vertical range of three cells. The function header will be Function CrossProd(Vect1 As Object, Vect2 As Object). Exercises 9 and 10 will help you get started.

4. Write a function to use the Newton–Raphson method to find the real roots of a cubic equation, one at a time. The worksheet should call the function using =Newton(guess, a, b, c, d) where

the arguments are the starting estimate and the coefficients of the quadratic equation. The main function should call other functions to compute $f(x)$ and $f'(x)$. In the loop of Newton, code an 'escape hatch' in the event that $f'(x)$ becomes equal to zero.

5. In Exercise 7 of Chapter 2 and Exercise 2 of Chapter 5 we developed a worksheet to find the equivalent resistance of four resistors in parallel. This provides an interesting example to demonstrate the For Each construct.

   Code the following user-defined function and use it in a worksheet similar to that below. The formula in E4 is =Resistor (A3:D3). This is copied down two rows. Note how the function can handle blank cells. Can you modify it to also handle zero values?

```
Function Resistor(List)
   Dim Item
   Resistor = 0
   For Each Item In List
          If WorksheetFunction.IsNumber(Item) Then
              Resistor = Resistor + 1 / Item
          End If
   Next Item
   Resistor = 1 / Resistor
End Function
```

	A	B	C	D	E
1	Resistors in Parallel				
2	r1	r2	r3	r4	R(equiv)
3	400	200	50	200	30.76923
4	10	10	10		3.333333
5	2	4			1.333333

9
Modelling I

Concepts

Frequently, physical models are used in experiments to predict the behaviour of the real thing. Thus a model aeroplane may be placed in a wind tunnel to test the design of a new aircraft. In these experiments, the behaviour of a *system* is being studied. The system here is the aircraft and the wind that is used to simulate movement of the aircraft through the atmosphere.

Mathematical models are another way of studying the behaviour of systems. You have often done this yourself. When we use $PV = nRT$ to solve a problem, we are assuming that the Ideal Gas Law is a good model for the behaviour of a gas. However, every model has its limitations. The ideal gas equation gives poor results at high pressures and low temperatures; the more complex model of van der Waals gives a more accurate description under these conditions. Similarly, Newton's laws give adequate results unless the velocities of the objects are moving at speeds approaching the speed of light; then we need to turn to relativity equations.

In this chapter we construct and analyse some simple models. Chapter 10 introduces Solver which may be used for models requiring optimization.

Exercise 1: Model of a Bouncing Ball

In this exercise we model the behaviour of a bouncing ball. We begin the design with a description of the system and the objectives of the model.

Description
A ball falling from a height, the surface upon which it bounces and the air through which it moves.

Objective
To generate data showing the maximum height reached by the ball after each bounce.

Now we need to state our assumptions, define the variables of the problem and state any known relationships between the variables. As we do this we may wish to refine the objectives.

Assumptions

While we would like the results to be accurate, there is a limit to the amount of work we are prepared to do for this problem. So our first assumption will be that air resistance may be ignored. Immediately, we have imposed a limit on the model. For a falling sphere, air resistance is negligible when the velocity is low. This means that our model is 'good' only when the initial height is not large. What other, unstated, assumptions have we made about the air?

Next we consider the surface. For simplicity, we make this a smooth, plane surface, perpendicular to the falling ball and rigidly attached to an object massively larger than the ball.

Now for the interaction of the ball and the surface. Consider a ball that drops from some height (h), hits the surface and bounces back to a new height (h'). In the idealized case of a perfect elastic collision $h = h'$. The theory of collisions states that:

Kinetic energy after collision = $k \times$ *KE before collision* (9.1)

where k is the coefficient of restitution with $k = 1$ for a perfectly elastic collision and $k < 1$ for inelastic collisions. Since the surface on which the ball falls is part of a massive object, we may ignore any changes in its kinetic energy and from Equation 9.1 derive:

$$mv^2_{\text{before}} = kmv^2_{\text{after}} \tag{9.2}$$

We are interested in the maximum heights reached by the ball after each bounce. Let v_{after} refer to the velocity as it starts its journey to reach height h_n – the height reached in the nth bounce. Then v_{before} will refer to the velocity of the ball falling from height h_{n-1}. From the conservation of energy:

$$mv^2_{\text{before}} = mgh_{n-1} \quad \text{and} \quad mv^2_{\text{after}} = mgh_n$$

We may therefore write Equation 9.2 as:

$$h_n = kh_{n-1} \tag{9.3}$$

We now see that our variables are: h_0, h_1, h_2, … and k and that Equation 9.3 relates these variables.

New objective

To generate data showing the maximum height reached by the ball after each bounce, and to observe how the behaviour changes with h_0 and k. The final worksheet will resemble Figure 9.1.

(a) Start a new worksheet. Enter the text and values in rows 1 to 4. The cell C2 holds the value of h_0; create the name *start* for this cell.

(b) Enter the text and values in row 5. These are the values of k we are using in the model.

(c) Enter the values 0 to 20 in column A. These number the bounces.

(d) In B6 enter the formula =start and copy this to B6:F6.

(e) In B7 enter the formula =B$5*B6 and copy this to B7:F7. Look at the formulas in this row. Note how they are in agreement with Equation 9.3.

(f) Copy B7:F7 down to row 26.

(g) We need to test our model. In B5 enter the values 0 and 1.0 successively and note the resulting h values. We should get $h_n = 0$ for $n > 0$ when $k = 0$ and $h_n = h_0$ for all n when $k = 1$. Replace the 0.9 value in B5 after this test.

	A	B	C	D	E	F
1	Bouncing Ball Model					
2		Start	20	cm		
3						
4	Bounce		Maximum height after each bounce			
5	k	0.9	0.8	0.7	0.6	0.5
6	0	20	20	20	20	20
7	1	18.00	16.00	14.00	12.00	10.00
8	2	16.20	12.80	9.80	7.20	5.00
9	3	14.58	10.24	6.86	4.32	2.50
10	4	13.12	8.19	4.80	2.59	1.25
11	5	11.81	6.55	3.36	1.56	0.63
12	6	10.63	5.24	2.35	0.93	0.31
25	19	2.70	0.29	0.02	0.00	0.00
26	20	2.43	0.23	0.02	0.00	0.00

Figure 9.1

(h) Make an XY chart of the data in A5:F26.

(i) The curves for each k value seem to be exponential. To check this add a trendline to the $k = 0.9$ curve using the options to display both the equation and the R^2 values on the chart. The R^2 value turns out to be 1.0 indicating a perfect fit.

(j) Add trendlines to the other curves with the display equation option only. The trendlines show the ball obeys the equation:

$$h_n = h_0 e(cn) \quad (9.4)$$

Make a note of the c value for each k value. If we examine Equation 9.3 we see that we could rewrite it as:

$$h_n = k^n h_0 \quad (9.5)$$

Clearly, the k and c values of Equations 9.4 and 9.5 must be related.

(k) On a blank part of the worksheet, show that the relationship is $c = \ln(k)$. Of course, now it is obvious: Equation 9.4 could be written as:

$$h_n = h_0 \exp(n \ln(k)) \quad (9.6)$$

(l) Save the workbook as CHAP9.XLS.

In this model we have observed how the behaviour of the system changes when we vary the value of k. This is called *parametric analysis.* We could also vary h_0. There are occasions when we are unsure of the exact value of a variable and make an educated guess at it. We might then change that parameter in small steps and observe how the behaviour of the system changes. This is called *sensitivity analysis.*

Exercise 2: Population Model

Description

An ecological niche contains two species; one is the prey, the other the predator. A theoretical analysis of the problem has yielded equations for the successive population of the two species:

$$N(t+1) = (1.0 - B(N(t) - 100))N(t)) - KN(t)P(t)$$
$$P(t) = QN(t)P(t)$$

where:

$N(t)$ = the population of prey in generation t
B = the net birth-rate factor
K = the kill-rate factor
$P(t)$ = the population of predators in generation t
Q = efficiency in use of prey

Objective

To observe how this model predicts the changing populations and to examine the sensitivity of the model to the values of the parameters.

(a) Open CHAP9.XLS and move to Sheet2. Enter the text and values shown in Figure 9.2. Select the range B6:F7 and use Insert|Name|Create to name the cells in B7:F7.

	A	B	C	D	E	F
1	Prey and Predator					
2						
3		Prey	N(t+1) = (1.0 - B(N(t) - 100) N(t) - KN(t)P(t)			
4		Predator	P(t+1) = QN(t)P(t)			
5						
6			No	Po	B	K	Q
7			50	0.2	0.005	0.5	0.0205

Figure 9.2

(b) We need some space for a graph, so move to A37 and enter the text shown in Figure 9.3. Fill the range A38:A78 with the series 1, 2, …, 40.

	A	B	C
37	t	N(t)	P(t)
38	0	50	0.20
39	1	57.50	0.21
40	2	63.83	0.24

Figure 9.3

(c) The formulas in B38:C39 are:
B38: =No
C38: =Po
B39: =(1 – B*(B38 – 100))*B38 – K*B38*C38
C39: =Q*B38*C38

(d) Copy B39:C39 down to row 78.

(e) In the space below row 8, create a chart from the data in A38:C78 making it 20 rows deep and 8 columns wide. The chart in Figure 9.4 uses two y-axes as discussed in Exercise 4 of Chapter 6.

The chart shows the classic prey/predator population behaviour: each species goes to a minimum before rising to a maximum but the two curves are shifted relative to each other. Note how the maximum population of each species is constant.

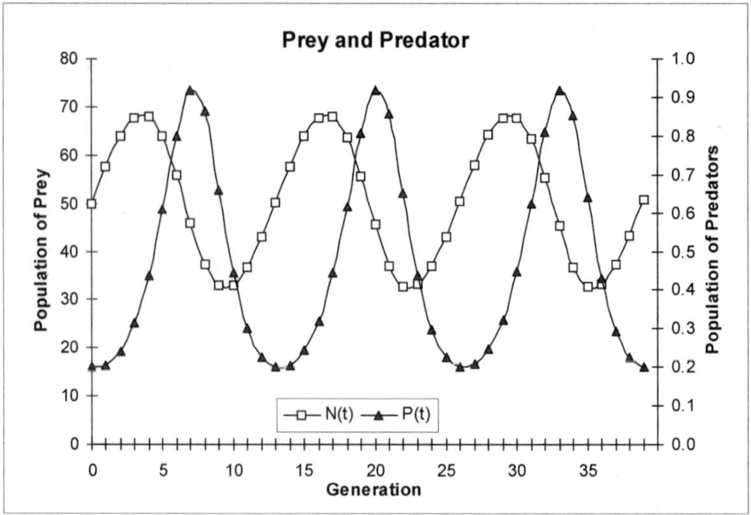

Figure 9.4

(f) We can now experiment with the parameters. After observing the effect of changing one parameter, re-enter its original value before changing the next.

(i) Change $N(0)$ to 60, 70, …
(ii) Change $P(0)$ to 0.25, 0.3, …
(iii) Change B to 0.0055, 0.0006, 0.00065, …
(iv) Change K to 0.25, 0.3, 0.19 ,0.18, …

Which parameters may be changed slightly while maintaining a stable state? Can you find one parameter for which a small change results in either a population explosion or an extinction?

Exercise 3: Titration Model

System
A volume of a weak acid (e.g. acetic acid) solution is placed in a vessel. Small quantities of a strong base solution (e.g. sodium hydroxide) are added.

Objective
To generate the data needed to create a chart showing how the pH varies with the volume of base added.

Variables

M_a the molarity (concentration) of the acid solution
V_a the volume of acid solution
K_a the dissociation constant of the acid
M_b the molarity (concentration) of the base solution
V_b the volume of base solution added
K_w the water constant 1.0×10^{-14}
pH the pH of the solution.

There will be several derived variables:
V_t total volume of the solution in the vessel
C_a concentration of acid remaining after V_b ml of base solution has been added
C_{ab} concentration of the weak acid's conjugate base
C_b concentration of unreacted base after equivalence point.

Relationships

Reaction of acid and base

$$HA + NaOH \rightarrow NaA + H_2O$$

Acid dissociation constant

$$K_a = \frac{[\text{conc of } A^-] \times [\text{conc of } H_3O^+]}{[\text{conc of remaining acid HA}]}$$

Water dissociation constant

$$K_w = [\text{conc of } H_3O^+] \times [\text{conc of } OH^-]$$

Definition of pH

$$pH = -\log_{10}[\text{conc of } H_3O^+]$$

Assumptions

There are three phases in this system and each has its own method of calculating the pH:

(i) *The amount of base added is less than the equivalence point.* If C_a is the concentration of the remaining acid, C_{cb} the concentration of A^- (the conjugate base of the acid), and x the concentration of H_3O^+, then:

$$\begin{array}{lcccccc} HA & + & H_2O & \rightleftharpoons & A^- & + & H_3O^+ \\ C_a - x & & & & C_{cb} + x & & x \end{array}$$

$$K_a = \frac{(C_{cb} - x)(x)}{(C_a - x)}$$

Solving the quadratic equation will give the concentration of H_3O^+ and hence the pH.

(ii) *The equivalence point* when the acid has been exactly neutralized by the base leaving a solution of NaA. If C_{cb} is the concentration of A^- and x the concentration of OH^-, then:

$$\begin{array}{llll} A^- & + \; H_2O & \rightleftharpoons HA & + \; OH^- \\ C_{cb} - x & & x & x \end{array}$$

$$K_b = \frac{K_w}{K_a} = \frac{(x)(x)}{(C_{cb} - x)}$$

Solving the quadratic equation will give the concentration OH^- from which the concentration of H_3O^+ and hence the pH may be found.

(iii) *After the equivalence point* when the pH is dependent only upon the concentration of unreacted base. If C_b is the concentration of OH^- formed from the unreacted base and x the concentration of H_3O^+ then K_w=[x][C_b] from which x is readily found.

We will differentiate the three cases with an IF function and use two user-defined functions to solve the quadratic equation in the first two steps. Note that these are required to return only one value: $\left(-b+\sqrt{b2-4ac}\right)/2a$.

(a) Open the VB Editor and insert a module on the CHAP9.XLS project. Code these two functions.

```
Function WeakAcid(CAcid, CConjBase, AcidConstant)
    b = CConjBase + AcidConstant
    c = -AcidConstant * CAcid
    WeakAcid = (-b + Sqr(b * b - 4 * c)) / 2
End Function

Function ConjugateBase(CConjBase, AcidConstant)
    Const Kw As Double = 0.00000000000001
    BaseConstant = Kw / AcidConstant
    b = BaseConstant
    c = -BaseConstant * CConjBase
    ConjugateBase = (-b + Sqr(b * b - 4 * c)) / 2
    ConjugateBase = Kw / ConjugateBase
End Function
```

Type the first statement in the ConjugateBase function as Const Kw = 1.0E–14. Visual Basic will change it to the form previously shown.

(b) Tab to Sheet3 and type the text and values shown in Figure 9.5 excluding C10. Click on the A column heading and use Format|Column|AutoFit Selection.

	A	B	C
1	Volumetric Titration of a Weak Acid		
2	Variables		
3	Molarity of Acid	Ma	0.025
4	Volume of Acid	Va	25
5	Acid constant	Ka	6.50E-05
6	Molarity of Base	Mb	0.025
7	Starting Base	Start	0
8	Aliquot amount	delta	1
9			
10	Moles of Acid	Xa	0.000625
11	Water constant	Kw	1.00E-14

Figure 9.5

(c) Select B3:C11 and use Insert|Name|Create to give C3:C11 the names in the corresponding B cells.

(d) Enter the formula =Ma*Va/1000 in C10.

(e) So we can put a chart near the variables, and hence be able to see the effect of changing them, our table will begin in K1. Enter the text from Figure 9.6 into K1:S1. Use [Alt] + [Enter↵] to make the text use more than one line in a cell.

(f) Type these formulas in row 2:
K2: =Start
L2: =Mb*K2/1000
M2: =(K2+Va)/1000
N2: =IF(ABS(Xa-L2) < 0.000001, "Equiv", IF(Xa>L2, "Acid", "Base"))

We might have used =IF(Xa=L2 ... in N2, but the use of an ABS function is safer because round-off errors may give values for Xa and L2 which are not quite equal when they should be.

O2: =IF(N2= "Acid", (Xa – L2)/M2, 0)
P2: =L2/M2
Q2: =IF(N2= "Base", (L2 – Xa)/M2, 0)
R2: =IF(N2="Equiv", ConjugateBase(P2,Ka), IF(N2="Base", Kw/Q2, WeakAcid(O2,P2,Ka)))
S2: =–LOG10(R2)

Format cells L2, M2, O2, P2 and Q2 to display 5 decimal digits, R2 to use scientific notation with two decimal places, and S2 to display 2 decimal digits.

	K	L	M	N	O	P	Q	R	S
1	Vol of Base ml	Moles of Base added	Total Volume litres	Test	Conc of Acid remaining	Conc of A ion	Conc of excess Base	H_3O^+	pH
2	0	0.000000	0.025000	Acid	0.02500	0.00000	0.00000	1.24E-03	2.91
3	1	0.000025	0.026000	Acid	0.02308	0.00096	0.00000	8.15E-04	3.09
4	2	0.000050	0.027000	Acid	0.02130	0.00185	0.00000	5.59E-04	3.25
5	3	0.000075	0.028000	Acid	0.01964	0.00268	0.00000	4.05E-04	3.39

Figure 9.6

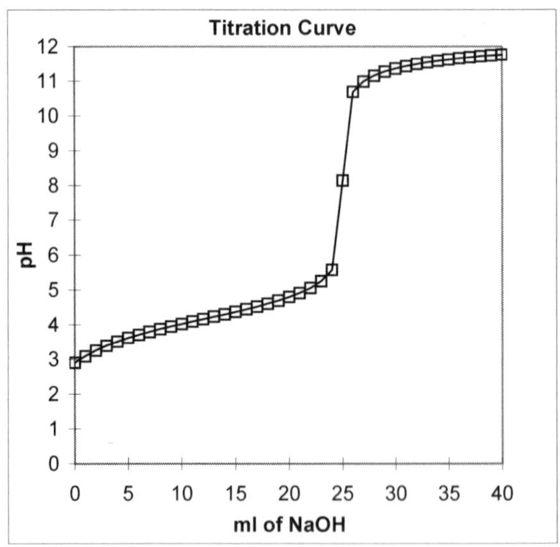

Figure 9.7

(g) To test the modules, formulas and variable values at this point, copy K2:S2 down to row 5 and enter the values shown in K3:K5 of Figure 9.6. Make any necessary corrections to formulas and user-defined functions.

(h) Change the formula in K3 to =K2+delta. Copy K3:S3 to row 42.

(i) Create an XY chart similar to Figure 9.7, using K2:K42 as the *x*-values and S2:S42 as the *y*-values.

(j) To see more detail over the range of interest use a value of 15 or 20 in C7. You can also change the value in C8 to get smaller increments of base added. Better yet, on either side of the equivalence point, use smaller increments to get a smoother curve.

(k) The next step is just for fun. Select S2:S42 and use the command Format|Cells|Number|Custom. In the Type box enter [Red][<7]0.00; [Blue][>7]0.00; [Green]0.00. This makes data in column S appear in red when the value is less than 7, blue when greater than 7 and green at 7. Notice how the values in the *y*-axis also use this format.

Exercise 4: Making Waves

In this exercise we demonstrate the use of Excel to explore the properties of waves. More specifically, we show how a Microsoft Excel chart may be used to show the effects of the superposition of waves. We shall use sine waves in the form $y = A \sin(2\pi ft)$, where A is the maximum amplitude, f the frequency and t the time.

It is important to use the correct sampling rate when charting sine waves. That is to say, the increment on the time axis should be small enough to faithfully capture a true sine wave. A useful rule-of-thumb is to use time increments of $1/12{*}f$ where f is the frequency of the wave. Anything much higher than this will fail to generate a sine wave. This is demonstrated in Figure 9.8.

(a) We begin by making a table of values for a sine wave with a small time increment and another with a larger increment. Cells in row 2 contain values except for the formulas:
D2: =1/(12*B2) This is the time increment
H2: =$C$2*SIN(2 * PI() * $B$2 * G2)
J2: =$C$2*SIN(2 * PI() * $B$2 * I2)

The formulas in row 3 are:
G3: =G2+$D$2
H3: =$C$2*SIN(2 * PI() * $B$2 * G3) Copied from H2
I3: =I2+$D$2*5
J3: =$C$2*SIN(2 * PI() * $B$2 * I3) Copied from J2

Copy G3:H3 down to row 60 and I3:J3 down to row 12.

	A	B	C	D	E	F	G	H	I	J
1		Frequency	Amplitude	Delta			Time	Wave A	Time	Wave B
2	Wave	440	1	0.000189			0	0	0	0
3							0.000189	0.500000	0.000947	0.500000
4							0.000379	0.866025	0.001894	-0.866025
5							0.000568	1.000000	0.002841	1.000000
6							0.000758	0.866025	0.003788	-0.866025
7							0.000947	0.500000	0.004735	0.500000
8							0.001136	0.000000	0.005682	0.000000
9							0.001326	-0.500000	0.006629	-0.500000
10							0.001515	-0.866025	0.007576	0.866025
11							0.001705	-1.000000	0.008523	-1.000000
12							0.001894	-0.866025	0.009470	0.866025
13							0.002083	-0.500000	0.010417	-0.500000
14							0.002273	0.000000		
15							0.002462	0.500000		
16							0.002652	0.866025		

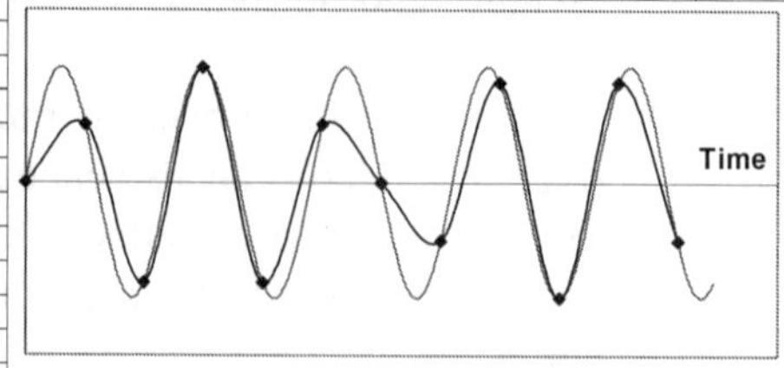

Figure 9.8

(b) Make an XY chart of the data in G2:H60. This data series is formatted to show no markers. Using the Copy and Paste Special technique introduced in Exercise 10 of Chapter 6 add I2:J12 as a second series. The two axes are formatted with *None* for the *Major tick marks*, *Minor tick marks* and *Tick mark labels*.

The data series made with the small time increment gives a fair representation of a sine wave. Something even smaller would be needed for a very smooth curve. The second data series is clearly unacceptable.

We next chart two sine waves having certain frequencies and amplitudes together with the sine wave that results from the superposition of the two – see Figure 9.9.

(c) We need a table for the two waves and for their superposition. Cells in row 2 contain values except for the formulas:

D2: =1/(12*B2)
E2: =MIN(D2:D3) This is our time increment
H2: =$C$2*SIN(2 * PI() * $B$2 * G2)
I2: =$C$3*SIN(2 * PI() * $B$3 * G2)
J2: =H2 + I2

The formulas in row 3 are:

G3: =G2+$E$2
H3: =$C$2*SIN(2 * PI() * $B$2 * G3) Copied from H2
I3: =$C$2*SIN(2 * PI() * $B$2 * G3) Copied from J2
J3: =H3 + I3

The formulas in G3:J3 are copied down to row 200.

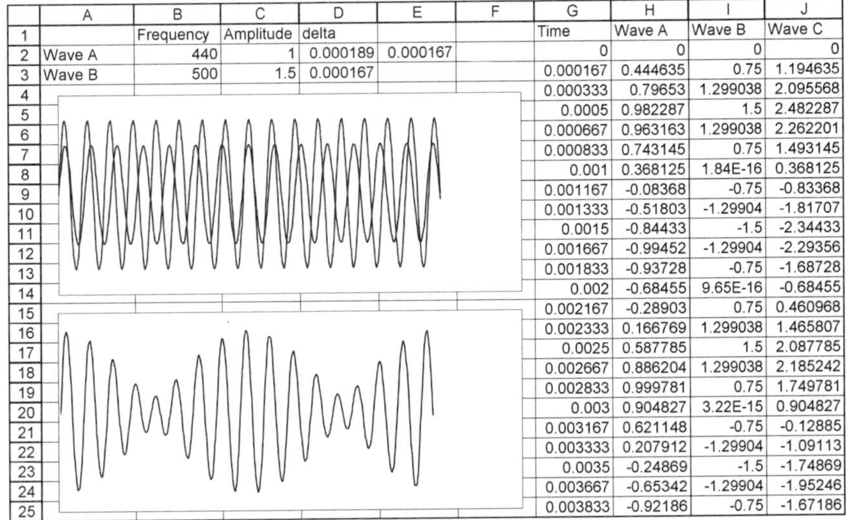

	A	B	C	D	E	F	G	H	I	J
1		Frequency	Amplitude	delta			Time	Wave A	Wave B	Wave C
2	Wave A	440	1	0.000189	0.000167		0	0	0	0
3	Wave B	500	1.5	0.000167			0.000167	0.444635	0.75	1.194635
4							0.000333	0.79653	1.299038	2.095568
5							0.0005	0.982287	1.5	2.482287
6							0.000667	0.963163	1.299038	2.262201
7							0.000833	0.743145	0.75	1.493145
8							0.001	0.368125	1.84E-16	0.368125
9							0.001167	-0.08368	-0.75	-0.83368
10							0.001333	-0.51803	-1.29904	-1.81707
11							0.0015	-0.84433	-1.5	-2.34433
12							0.001667	-0.99452	-1.29904	-2.29356
13							0.001833	-0.93728	-0.75	-1.68728
14							0.002	-0.68455	9.65E-16	-0.68455
15							0.002167	-0.28903	0.75	0.460968
16							0.002333	0.166769	1.299038	1.465807
17							0.0025	0.587785	1.5	2.087785
18							0.002667	0.886204	1.299038	2.185242
19							0.002833	0.999781	0.75	1.749781
20							0.003	0.904827	3.22E-15	0.904827
21							0.003167	0.621148	-0.75	-0.12885
22							0.003333	0.207912	-1.29904	-1.09113
23							0.0035	-0.24869	-1.5	-1.74869
24							0.003667	-0.65342	-1.29904	-1.95246
25							0.003833	-0.92186	-0.75	-1.67186

Figure 9.9

(d) The first chart is constructed from the data in G2:I200. The second uses G2:G200 as the *x*-values and J2:J200 for the *y*-values. The two axes and the plot area have been removed.

The lower chart clearly shows the development of beats when two sound waves of similar frequency interfere. The techniques of this exercise may be expanded to experiment with three or more waves.

Exercise 5: Taking Control

We can perform many virtual experiments using the worksheet developed in Exercise 4 by changing the frequencies and the amplitudes of the waves. In this exercise we take a brief look at some of the more advanced features of Microsoft Excel that make it more convenient to change these values. We will construct a worksheet containing two controls (a scroll bar and a spinner) and which will resemble that in Figure 9.10.

(a) We need to rearrange the worksheet to make room for the controls. Select the column heading G:L, right click and select the *Insert* option. Drag the two charts into the space you have created.

(b) Use View|Toolbars to display the *Forms* toolbar. By allowing the pointer to linger on each icon on the toolbar and display a tool tip, locate the *Scrollbar* and the *Spinner* tools.

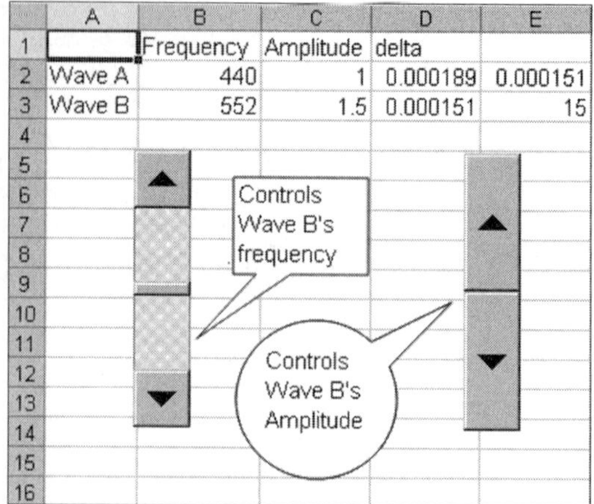

Figure 9.10

(c) Click on the Scrollbar tool and observe that the mouse pointer changes to a closed cross (+) when over a cell. Drag this symbol to outline the shape of the scrollbar.

(d) Right click on the scrollbar you have just made and select the *Format Control* item from the popup menu. On the *Control* tab of the *Format Control* dialog box (Figure 9.11), set the minimum value to 100, the maximum to 1000 and cell link to B3. Close the dialog box.

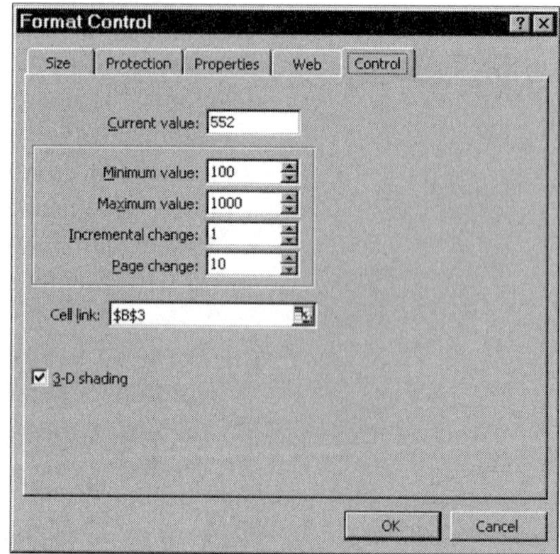

Figure 9.11

(e) Note that moving the slider part of the scroll changes the value in B3. Clicking on the arrows on the scrollbar changes B3's value by 1 unit (this is the *increment change* value on the dialog box) while clicking above or below the slider changes the value by 10 units (this is the *page change* value.)

(f) Using a similar method, add a spinner to the worksheet. We would like to be able to change C3's value from 0 to 5 in 0.1 increments but the spinner permits only integer increments. We can solve this by linking the spinner to another cell (E3) and making the maximum value 50. In C3 enter the formula =E3/10. Thus a change of 1 unit in E3, caused by using the spinner, results in a change of 0.1 in C3's value.

(g) We can add some documentation to the worksheet. Use View|Toolbars to display the *Drawing* toolbar. Experiment with the *AutoShapes* to add the two 'callouts' shown in Figure 9.10.

The tools on the Forms toolbar are actually a little outdated and have been replaced by those on the Control toolbar. These are called *ActiveX* controls. However, they are somewhat more complicated to set up. If you wish to modify a control, the Control toolbar must be visible and you must be in Design Mode – toggle the icon depicting a protractor. To link a control to a cell use the Properties item on the menu that pops up when you right click a control. Unlike control from the Forms toolbar, ActiveX controls may be given colours. They can also have VBA code attached to them – a topic beyond the scope of this book.

Problems

1.* A recycling plant has a number of one-litre bottles which originally contained a solution of a toxic chemical X. The bottles are nominally empty but 1 millilitre of solution remains in each. The washing process for each bottle is: add V millilitres of water, stir the bottle, drain the solution into a vat. We will assume that 1 millilitre of solution remains after draining. This process is repeated until the concentration of X in the remaining solution is less than some required value. There is a labour cost associated with each washing step and a cost to dispose of the solution in the vat. Your task is to find the optimum value of V taking into account these costs, and the initial and required concentrations.

(a) Start with a worksheet similar to that shown in Figure 9.12. The formula in the B column either computes the new concentration or returns #N/A if no further washes are needed. Experiment with various values for *vol* (the amount of water added in each washing step) to find how to achieve a final concentration less than or equal to that required at minimum cost.

	A	B	C	D	E
1	Bottle Washing Problem				
2					
3	Initial concentration		init	0.5	
4	Required conc		req	1.00E-08	
5	Cost per wash		wash	0.25	per wash
6	Cost of disposal		disp	1.50	per litre
7					
8	vol	750			
9	Washes	Conc	Cost		
10	0	0.5	0		
11	1	0.000666	1.375		
12	2	8.87E-07	2.750		
13	3	1.18E-09	4.125		
14	4	#N/A	#N/A		

Figure 9.12

(b) Improve your worksheet to give results similar to those in Figure 9.13.

	A	B	C
8	Volume	Washes	Cost
9	5	10	2.58
10	10	8	2.12
11	20	6	1.68
12	30	6	1.77
13	40	5	1.55
14	50	5	1.63
15	60	5	1.70
16	70	5	1.78
17	80	5	1.85
18	90	4	1.54
19	100	4	1.60
20	110	4	1.66
21	120	4	1.72
22	130	4	1.78

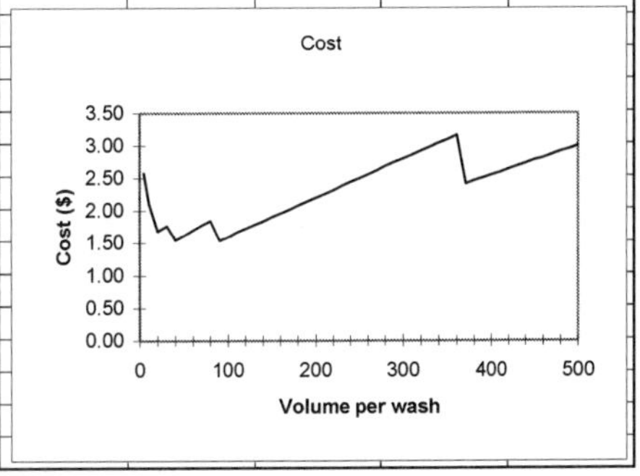

Figure 9.13

2.* The radioactive decay sequence shown in Equation 9.7 occurs in nuclear reactors. When the reactor is operating the neutron flux destroys the I and Xe. When it is shut down there is a residual concentration of each isotope. Because the half-life of I^{135} is smaller than that of Xe^{135}, the concentration of the latter reaches a maximum and then decays to zero. The reactor cannot be restarted until the Xe^{135} is well passed its maximum. The equations governing the production of the two isotopes are:

$$\mathrm{I}^{135} \xrightarrow{6.68\mathrm{hrs}} \mathrm{Xe}^{135} \xrightarrow{9.13\mathrm{hrs}} \mathrm{Cs}^{135}$$

$$\frac{d[\mathrm{I}]}{dt} = -k_{\mathrm{I}}[\mathrm{I}]$$

$$\frac{d[\mathrm{Xe}]}{dt} = -k_{\mathrm{Xe}}[\mathrm{Xe}] + k_{\mathrm{I}}[\mathrm{I}] \qquad (9.7)$$

where [Xe] and [I] denote concentrations, and k_{Xe} and k_{I} are the decay constants. The decay constant k of a radioisotope is related to its half-life λ by $k\lambda = \ln 2$.

Your task is to model this system and show how the concentration of Xe varies with time for given initial concentrations of I and Xe. We will approximate the first equation in Equation 9.6 as $\Delta[\mathrm{I}] = -k_{\mathrm{I}}[\mathrm{I}]\Delta t$, giving $[\mathrm{I}]_t = [\mathrm{I}]_0(1 - kt)$, where $[\mathrm{I}]_0$ is the initial concentration of I^{135} when the reactor is shut down, and $[\mathrm{I}]_t$ is the concentration after time t. What condition is needed for this approximation to be justified? The equation for [Xe] is treated similarly. Construct a worksheet similar to that in the figure below. Plot the data A7:C108. Experiment with the values in D3:D5 to observe the behaviour of the model.

	A	B	C	D
1	Reactor Problem			
2		half-life (hrs)	k (hr^{-1})	Initial conc
3	Iodine	6.68	0.1038	2
4	Xenon	9.13	0.0759	0.001
5		Time interval		0.5
6				
7	t	I conc	Xe conc	
8	0.00	2	0.0010	
9	0.50	1.8962	0.1047	
10	1.00	1.7979	0.1991	
100	46.00	0.0149	0.1565	
101	46.50	0.0141	0.1513	

3.* Because microprocessors have limited memory, their programs must be kept very small. The algorithm shown below has been suggested as a quick way to generate two cycles of a sine wave. The value of Quick(1) is an approximation to sin(90); Quick(n) approximates $\sin(90-5.625*(n-1))$.

```
Start with q(1) = 128 and d(1) = -1
Quick(1) = q(1)/128
For n = 2 to 129
     d(n)          = d(n-1) - 1     when q(n) >= 0
                   = d(n-1) + 1     when q(n) < 0
     q(n)          = q(n-1) + d(n)
     Quick(n)      = q(n)/128
Next n
```

Your task is to compare the results from this algorithm with the true sine values. The figure below shows how to start the worksheet. Carefully consider the entries needed in row 3 which will allow you to copy that row down to row 130. Plot the data in the Quick and Sine columns against that in the n column.

	A	B	C	D	E	F
1	n	d	q	Quick	Angle	Sine
2	1	-1	128	1	90.000	1
3	2	-2	126	0.9844	84.375	0.9952
4	3	-3	123	0.9609	78.750	0.9808

10
Solving Equations

Concepts A: Finding Roots

In this section we examine methods of finding roots of non-linear equations such as polynomial ($3x^3 - 7x^2 - 22x + 40 = 0$) and transcendental ($\exp(-x) - \sin(x) = 0$). If the equation is written as $f(x)$ then a *root* of the equation is a value of x such that $f(x) = 0$. The value of x is sometimes called the zero of the function. Some equations may be solved analytically. The quadratic formula, for example, is used to find the roots of a quadratic equation. With other equations the analytical method may be very complex or not exist at all. In these cases we may use numerical methods to find approximate roots. One should also remember the usefulness of graphing a function to determine the number and values of its roots.

Microsoft Excel includes two tools (Goal Seek and Solver) for finding roots. A discussion of the algorithms used by these tools is beyond the scope of this book but if you are familiar with the bisection or the Newton–Raphson method you will have some appreciation of how they work. We show in the first exercise how the bisection, or interval halving, method may be implemented on a worksheet. It is left as an exercise to the interested reader to develop a worksheet implementation of the Newton–Raphson method. Subsequent exercises use Goal Seek and Solver to find approximate roots.

Exercise 1: The Bisection Method

In Figure 10.1 the values of $F(a)$ and $F(b)$ lie on opposite sides of the x-axis. Therefore there is a root of $F(x)$ lying between a and b. Let m be the midpoint of the interval a to b. Since $F(m)$ has the opposite sign of $F(b)$, this root lies between m and b. By halving the interval we have a more accurate idea of the value of the root. Looking at the function $G(x)$ we see that the root lies between a and m. So we must use the values m and a to find the next approximation. Of course we may repeat this halving over and over; successive iterations giving smaller intervals – the a and b values will converge.

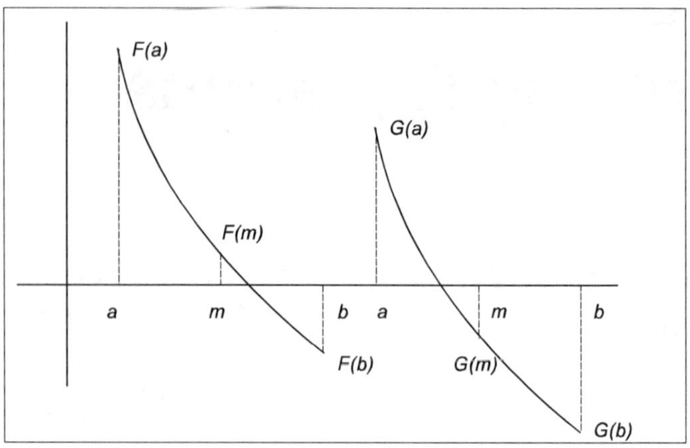

Figure 10.1

This allows us to develop an algorithm for finding a root of $f(x)$:

Start with values of a and b such that $f(a)$ and $f(b)$ have opposite signs
Loop until the required accuracy is achieved
    Find the midpoint $m = (a + b)/2$
    If $f(m)$ and $f(b)$ have opposite signs
            give a the value of m
    Else
            give b the value of m
    End if
End loop.

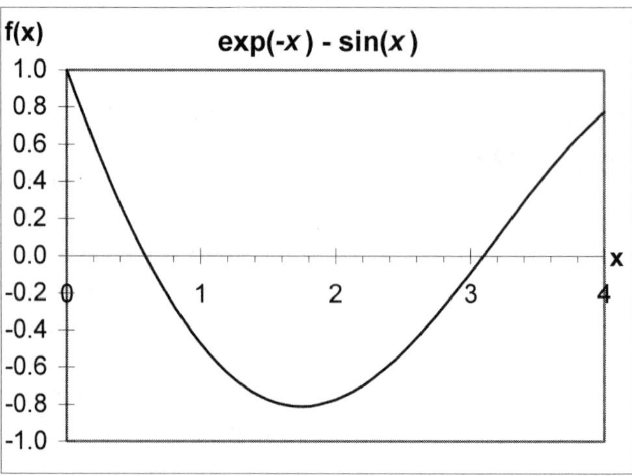

Figure 10.2

To demonstrate how we may implement this algorithm in Excel, we shall find the roots of the function exp(–*x*) – sin(*x*). Figure 10.2 shows a plot of this function for values of *x* from 0 to 4.

Clearly, this equation has one root at approximately 0.6 and another near 3. Our task is to find more exactly what these values are. In subsequent exercises we use Goal Seek and Solver to find the roots of this equation and compare their results with those obtained in this exercise.

(a) Open a new workbook. On Sheet1 enter the text shown in A1:F3 of Figure 10.3.

(b) On row 4 enter:

A4:	0.5	The first *a* value
B4:	1	The first *b* value
C4:	=(A4+B4)/2	Compute the midpoint *m*
D4:	=EXP(–A4) – SIN(A4)	The value of *f*(*a*)
E4:	=EXP(–B4) – SIN(B4)	The value of *f*(*b*)
F4:	=EXP(–C4) – SIN(C4)	The value of *f*(*m*)

To save time, the formula in D4 may be copied to E4:F4 by dragging D4s fill handle to the right two cells.

Row 4 sets the initial conditions. Next we compute the next interval. In Row 5 we compute the first approximation.

(c) In A5 enter the formula = IF(SIGN(F4)<>SIGN(E4), C4, A4). This compares the signs of *f*(*m*) and *f*(*b*). If they differ then cell A5 (the new *a* value) is given the value of *m* of the first approximation. Otherwise the cell retains the old *a* value.

	A	B	C	D	E	F
1	Bisection method					
2						
3	a	b	midpoint	f(a)	f(b)	f(midpoint)
4	0.5	1	0.75	0.127105	-0.47359	-0.20927
5	0.5	0.75	0.625	0.127105	-0.20927	-0.04984
6	0.5	0.625	0.5625	0.127105	-0.04984	0.03648
7	0.5625	0.625	0.59375	0.03648	-0.04984	-0.00722
8	0.5625	0.59375	0.578125	0.03648	-0.00722	0.014495
9	0.578125	0.59375	0.585938	0.014495	-0.00722	0.003603
21	0.588531	0.588535	0.588533	1.73E-06	-3.6E-06	-9.1E-07
22	0.588531	0.588533	0.588532	1.73E-06	-9.1E-07	4.11E-07
23	0.588532	0.588533	0.588533	4.11E-07	-9.1E-07	-2.5E-07
24	0.588532	0.588533	0.588533	4.11E-07	-2.5E-07	8.01E-08

Figure 10.3

(d) In B5 enter the formula =IF(SIGN(F4)<>SIGN(E4),B4,C4). This keeps the old value for b when the signs of $f(m)$ and $f(b)$ differ but uses the old m value for the next b when the signs are the same. The values in A5 and A4 are equal when a was not replaced by m; in which case the new b value is the m value from the first approximation. Otherwise the previous b value is used.

To compute successive iterations we copy row 5 down the sheet. But for how may rows? Recalling that each interation halves the interval, we note that 20 iterations will reduce the interval by a factor of 2^{20} or about a millionfold. Surely this will be more than enough!

(e) Copy C4:F4 down to row 24. In Figure 10.3 rows 10:20 have been hidden to make the figure smaller.

In row 4 $f(b)$ and $f(m)$ have the same sign, so the new b value in row 5 is the previous m value. The same occurs when going from row 5 to row 6. But now $f(b)$ and $f(m)$ have the same sign, so in row 7 the m value is passed to a.

On row 24 with $x = 0.588533$, the function evaluates to 8×10^{-8} which is acceptably close to 0. The values in the A and B columns are not changing very much at this point. You may wish to copy row 24 down to row 50. At this point the function evaluates to approximately 1×10^{-15} so we are at the limit of precision of Excel. You will not see any changes in the a and b values unless you widen the columns or use a formula to display the difference in successive values.

(f) From Figure 10.2 we know there is a root near 3. Replace the initial values of a and b in line 4 to find this second root. It does not matter much if you use 3 and 4, or 3 and 3.5. Why is this?

(g) Save the workbook as CHAP10.XLS.

Finding Roots with Goal Seek

How would you answer this question: For what value of x does the function $3x^3 - 10x^2 - x + 1$ evaluate to 100? You could find the answer by trial and error. Enter some value for x in A1 and in B1 enter the formula = 3*A1^3 – 10*A1^2 – A1 + 1. Now vary A1 until the desired result is obtained. This is exactly what Excel's Goal Seek does but with the help of a mathematical algorithm.

When Goal Seek is running you specify three things: (i) that B1 is the cell of interest – the *Set cell*, (ii) that the value you require is 100 – the *To value,* and (iii) that A1 is the cell whose value is to be changed – the *By changing cell.* Goal Seek changes the value of A1 until it finds a value which gives the formula in B1 the value close to 100. If you had specified a value of 0 rather than 100, then the value in A1 would be one of the roots of the functions in B1.

Goal Seek is a very easy tool to use but it has its limitations. In the next section of this chapter we see that Solver is far more powerful.

Exercise 2: A Simple Quadratic Equation

In this exercise we will find the roots of $2x^2 - 5x - 12 = 0$ using Goal Seek. The plot in Figure 10.4 will help us understand which solution Microsoft Excel finds. If we make an initial guess of 0, denoted by point G_1 on the plot, then Goal Seek will find the root with value −1.5 at the point R_1. Goal Seek 'explores' the point G_1 and determines that the function moves closer to zero as x becomes more negative. Conversely, if the initial guess is 3 (the point G_2), Goal Seek finds the root with the value 4.

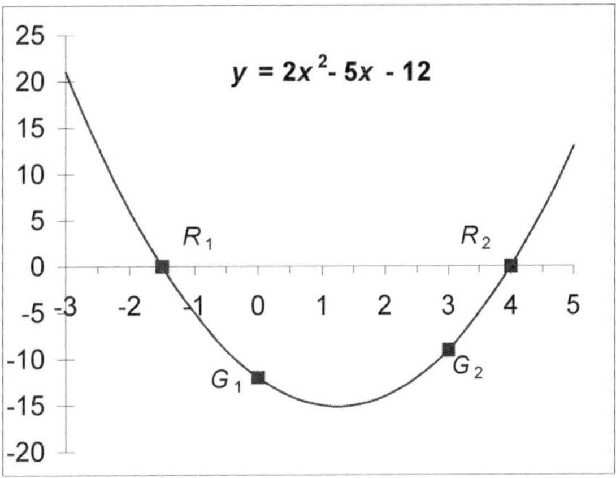

Figure 10.4

(a) Open the workbook CHAP10.XLS and move to Sheet2. Start with a worksheet similar to that in Figure 10.5. In B3 type the formula =2*A3*A3–5*A3–12. Copy this to cell B4.

(b) Make B3 the active cell. On the menu bar, click Tools followed by Goal Seek. Complete the Goal Seek dialog box as shown in Figure 10.6. You may type A3 in the *By changing cell* box, or, with the box selected, click on the A3 cell. In the second case, Excel enters the value $A$3. Now click the OK button.

	A	B	C
1	To solve $2x^2 - 5x - 12 = 0$		
2	Root	Function	
3	0	-12	
4	3	-9	

Figure 10.5

Goal Seek

Set cell: B3

To value: 0

By changing cell: $A$3

OK Cancel

Figure 10.6

(c) The Goal Seek Status dialog box appears; see Figure 10.7. This reports that our target value was 0 and Goal Seek has obtained a value of –3.3424E – 05. This is very close to zero so we will click the OK button. Goal Seek has changed the value of A3 to –1.5. With this value of *x* the function evaluates essentially to zero (-3.3×10^{-5}) so this value is a root of the function.

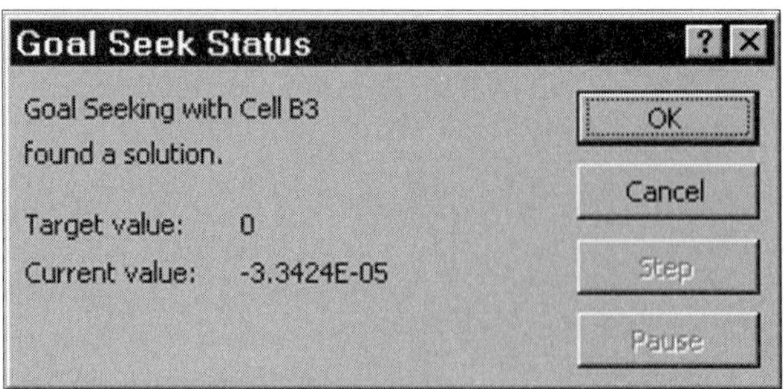

Figure 10.7

(d) Make B4 the active cell and repeat steps (b) and (c) to find the next root. On my PC the value in A4 becomes 4 and B4 has a value of –2.5E – 06. Save the workbook CHAP10.XLS.

Accuracy

Why does the worksheet report values for the function that are not exactly zero when it uses *x* values that appear exactly correct for the roots? If you widen the A column the answer is that Goal Seek did not find the *exact* value –1.5 and 4. My PC gave the values –1.49999696145665 and 3.99999977326809, respectively. Goal Seek uses an iterative algorithm to get closer and closer to the solution. It therefore needs to stop at some point. In our case it stopped just short of the exact solutions. Type the values –1.5 and 4 into cells A3 and A4, respectively. The two function values will now be exactly 0.

The problem with using Goal Seek or Solver to find the roots of quadratic equations is that you have to provide an initial guess. If the equation has one real root you will generally have no problem finding it. When there are two roots, your initial guesses may all converge to the same solution. This frustration can be avoided by using a worksheet based on the quadratic formula as demonstrated in an earlier chapter.

Exercise 3: Solving a Cubic Equation

If we have only one quadratic equation to solve it is probably more efficient to use the quadratic formula manually rather than setting up a worksheet. Cubic equations are a different matter. Here the tasks of trying various guesses is worth the effort. When finding the solution to a cubic equation is part of a physical problem, we may know the approximate value of the root in which we are interested or there may be only one real root. Either of these cases will simplify the task of making the initial guess.

In this exercise we will set up a worksheet that may be used to solve a cubic equation. We shall used *named* cells. You should recall from an earlier exercise that if we attempt to use 'c' as a name Microsoft Excel replaces this by 'c_'.

(a) Open CHAP10.XLS and on Sheet3 enter the values of all the cells except E4 to E6 as shown in Figure 10.8.

(b) Select the range A4:B7 and use Insert|Name|Create to name the cells B4:B7 as 'a', 'b', 'c_', and 'd', respectively. Note that with the values shown in B4:B7, we have set the worksheet to solve the equation $2x^3 + x^2 - 246x + 360 = 0$. When you typed 'c' in A6, did Excel change it to 'Coefficients'? Use Ctrl+z to undo the change. If you find the *AutoComplete* feature annoying, turn it off on the *Edit* tab of Tools|Options.

	A	B	C	D	E
1	Cubic equation solver				
2					
3	Coefficients			Roots	Function
4	a	2		-20	-10320
5	b	1		0	360
6	c	-246		20	11840
7	d	360			

Figure 10.8

(c) The general expression for a quadratic function is $f(x) = ax^3 + bx^2 + cx + d$. In E4 type the formula =a*D4^3 + b*D4^2 + c_*D4 + d. If Excel reports 'Error in formula', check that you typed 'c_' not 'c'. Copy this to cells E5 and E6. Have you remembered the shortcut way to do this – clicking on the fill handle of E4?

Now we are ready to use the worksheet. Note that the starting values shown in D4:D6 are not quite arbitrary; they have been chosen to give the reader three roots to the function. In 'real' cases, the users will need to experiment a little to find satisfactory starting values.

(d) Move to E4 and use Goal Seek to find the first solution by varying D4 to give E4 as a zero value.

(e) Move to E5 and use Goal Seek to find the first solution by varying D5 to give E5 as a zero value.

(f) Repeat step (e) with cells E6 and D6. Cells D4:D6 should now have the three solutions –12, 1.5 and 10. Of course, Goal Seek will not give these values exactly but you can discover that these are the exact solutions.

(g) Test your understanding of the process by finding the solutions of $3x^3 - 12x^2 - 255x + 1120 = 0$. One root is approximately 5, the others lie on each side of this root. Save the workbook CHAP10.XLS.

Exercise 4: Transcendental Equations

Goal Seek may be used to solve transcendental equations. The first equation we solve in this exercise is the same as that solved in Exercise 1 using the bisection method.

(a) Open CHAP10.XLS. On Sheet4 enter the text in A1:A4 and B2:C2, and the values in B3:B4 as shown in Figure 10.9.

	A	B	C
1	Goal Seek and Transcendental equations		
2		x	f(x)
3	Exp(-x) - Sin(x)	0	1
4	Cos(x) -Tan(x)/2	1	-0.2384

Figure 10.9

(b) Enter the formulas:
C3: =EXP(–B3) – SIN(B3)
C4: =COS(B4) – TAN(B4)/2

(c) Make C3 the active cell and call up Goal Seek from the Tools menu. The *Set cell* is C3, the *To value* is 0, and the *By changing cell* is B3. Click OK. How does the result compare with that obtained in Exercise 1?

(d) Find the root of $\exp(-x) - \sin(x) = 0$ with a value close to 3.

(e) Find two positive roots for $\cos(\theta) - \tan(\theta)/2 = 0$.

(f) Save the workbook CHAP10.XLS.

Using Excel's Solver

The Solver Add-In is much more powerful than Goal Seek. It was originally designed for optimization problems (problems that are the realm of operational research experts) but it is useful for root finding and similar mathematical problems. It differs from Goal Seek in a number of significant ways. Some of these are:

Note: In each Excel session, when you first call up Solver with Tools|Solver it is normal for Solver to take some time to load. Subsequent call-ups will respond much faster.

(i) When you have used Solver once on a worksheet, it will retain its settings when it is next used on that worksheet.

(ii) It is possible to save one or more 'models'. We will not pursue this topic.

(iii) Whereas Goal Seek allows you to vary one cell, with Solver you can vary 200 cells but using no more than 16 ranges.

We could vary, for example, A1:A10 and B1.

(iv) Solver permits constraints. For example, you can require that a varied cell always has a positive value.

(v) Solver may be used to find the value of the variables that give the formula a maximum or a minimum value as well as a set numeric value.

(vi) We may control how Solver finds a solution. See Solver Options below.

Solver should be found as one of the items on the Tools menu. If it is missing try using Tools|Add-Ins. Failing this you will need to reinstall Excel specifying that you require Solver to be installed; then use Tools|Add-Ins to load it.

Solver is licensed to Microsoft by Frontline Systems, Inc. whose web site (www.solver.com) has much valuable information on the product, as has the Excel Help facility.

Exercise 5: Roots of a Cubic Equation with Solver

In this exercise we use Solver to find the roots of the cubic equation we investigated in Exercise 3. As with Goal Seek, when a function has many roots, Solver will locate the one closest to the starting value (sometimes called the *guess*).

(a) Open the workbook CHAP10.XLS and insert a new worksheet – Sheet5. Move to Sheet3, select A1:E7 and click the Copy button. Move to Sheet5 and, with A1 as the active cell, click the Paste button.

(b) Select A4:B7 and use Insert|Name|Create to name B4:B7, otherwise your formulas will refer to cells on Sheet3. Reset the values of D4:D6 to –20, 0 and 20, respectively. Your worksheet should now resemble that in Figure 10.8.

(c) Move to the cell E4 and select Solver from the Tools menu. The Solver dialog box appears – Figure 10.10.

(d) Ensure that the *Set Target Cell* box contains the reference $E$4, that the *Value of* radio button is selected and the text box contains the value 0.

(e) Use the mouse to move to the *By Changing Cells* box. Either type 'D4' in this cell (it will change to $D$4) or use the mouse to click on the cell D4.

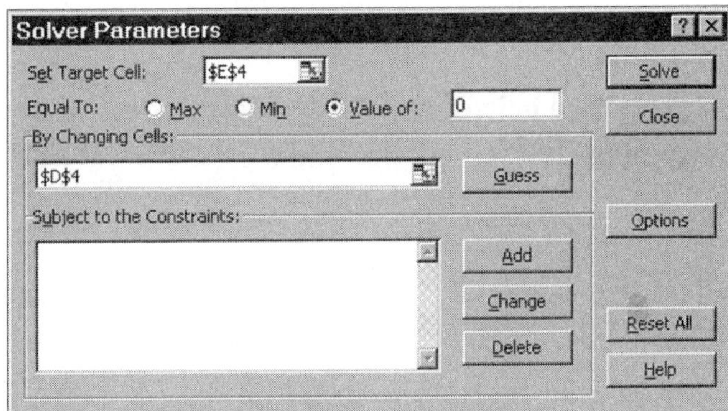

Figure 10.10

(f) Click on the *Solve* button. After a second or two, Solver will report whether or not it has found a solution; see Figure 10.11. Click the 'OK' button. With a starting value of −20, your first solution should be −12.

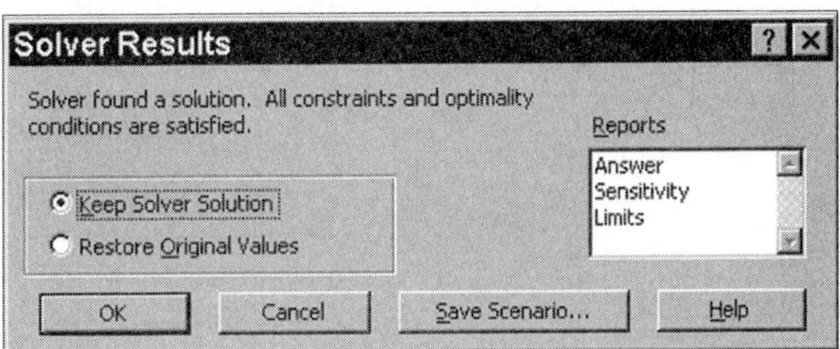

Figure 10.11

(g) Repeat steps (c) to (f) with E5 as the *Set Target Cell* and D5 as the *By Changing Cell* to find the second root of the cubic equation.

(h) Repeat steps (c) to (f) with E6 as the *Set Target Cell* and D6 as the *By Changing Cell* to find the third root of the cubic equation.

Cell D6 should now display 10 but the formula bar will show its actual value is not exactly this. Enter a value of 50 in D6 and call up Solver again. This time you may get exactly 10 for the answer. Solver uses a series of approximations to get its solution so it is not surprising that the final result depends on the starting value. If you are getting very different results, read the note on Solver Options later in this chapter.

(i) Save the workbook.

Exercise 6: Using a Constraint

The ability to set constraints is essential in optimization problems but less so with the types of problems we are solving. However, for demonstration purposes, we shall look at a simple example of their use when solving a cubic equation. Suppose we have to find a root for the equation: $1.3x^3 - 2.45x^2 - 0.8x + 1.25 = 0$. Let us further suppose that the problem that gave rise to this equation tells us that the value of x which interests us lies between 1 and 2.

(a) Open CHAP10.XLS and on Sheet6 set up a worksheet to look similar to that in Figure 10.12. Name the cells B4:B7. The only formula is in E4; it is =p*D4^3+q*D4^2+r_*D4+s. Note that we use 'r_' not 'r' for the cell C6 since Microsoft Excel reserves the names 'r' and 'c' for its own use. The value of 1.5 in D4 is our initial guess at the root.

	A	B	C	D	E
1	Finding a positive root of a cubic equation				
2					
3	Coefficients			Roots	Function
4	p	1.3		1.5	-1.075
5	q	-2.45			
6	r	-0.8			
7	s	1.25			

Figure 10.12

(b) Call up Solver as before. Set the *Target Cell* to E4 and specify zero for required value.

We will now add two constraints. We will specify that D4 is to be greater than 1 and less than 2.

(c) Click on the *Add* button in the *Subject to the Constraints* area to bring up the Add Constraint dialog box shown in Figure 10.13. In the *Cell Reference* box type D4. Change the operator to '>=', and type '1' in the Constraint box.

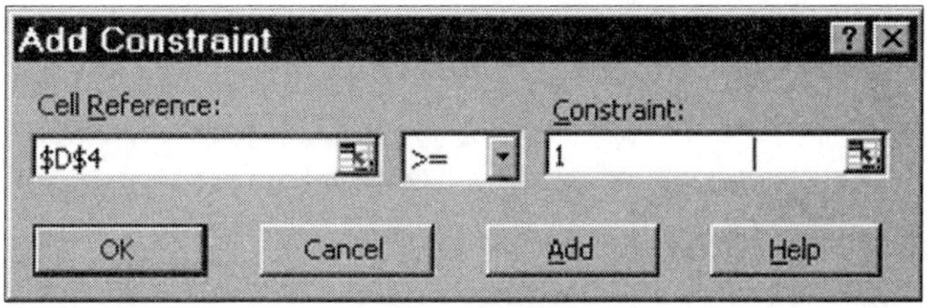

Figure 10.13

Click on the *Add* button of the Add Constraint dialog to add the second constraint. Make this read D4 <= 2. Then click on the *OK* button since we have no more constraints to add. Figure 10.14 shows the Solver dialog with two constraints.

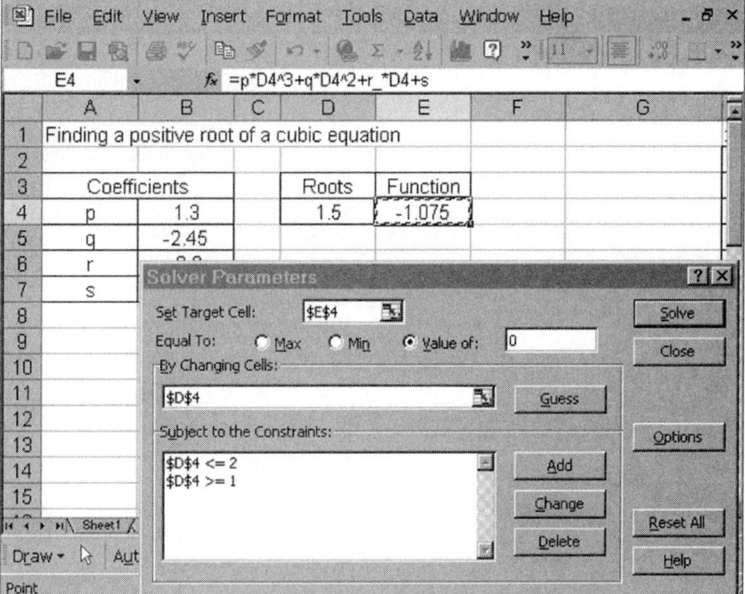

Figure 10.14

(d) Click the *Solve* button. Solver finds an acceptable result. My values are D4 = 1.94703739312482... and E4 = 7.87E–07.

(e) Change the constraint to find the other two values. One of them is negative so the constraint D4 <= 0 will be appropriate. The other lies between 0 and 1. With a starting value of 0 it may find a solution with no constraints. Good hunting!

(f) Save the workbook.

Solver Options

We shall have little need in these Exercises to alter Solver's operational values but the reader may be interested in looking at them. Click on the Solver dialog *Options* button to open the Options dialog shown in Figure 10.15.

Note:. Avoid having IF or CHOOSE functions between the decision variables (the ones in By Changing) and the objective (the Target Cell) since this will generally cause Solver to fail to find a solution.

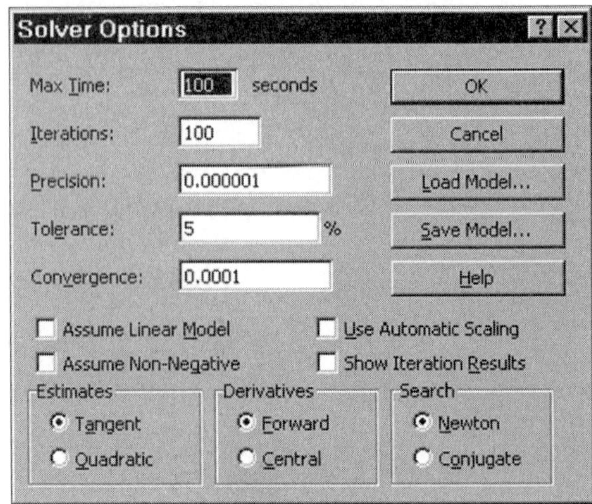

Figure 10.15

Max Time sets the maximum amount of time Solver may spend on the problem. The default value of 100 seconds is ample with a modern PC for all but very large problems.

Iterations sets the limit of the number of attempts Solver has to find a satisfactory solution.

Precision pertains to the constraints. Let the Precision be 1×10^{-6} and suppose we specify the constraint A1 >= 0. After some iterations Solver find a solution but A1 = -1×10^{-7} . Solver will consider the constraint has been met since it is within the precision.

Tolerance pertains to integer constraints. An integer constraint makes the problem much harder. Try initially solving without integer constraints.

Convergence sets the amount of relative change to allow in the last five iterations before Solver stops with a solution.

Assume Linear Model determines which algorithm is used by Solver. Linear problems are more readily solved.

Use Automatic Scaling is too technical to explain here.

Assume Non-Negative ensures that all decision variables (the ones in By Changing) that are not explicitly given a constraint have a lower bound of zero.

Explanations of the other options are beyond the scope of this book. However, as we will see in later Exercises, setting Derivatives to Central can be advantageous when solving for roots of equations.

Concepts B: Solving Simultaneous Equations

A system of linear equations is a set of n linear equations in which each equation contains up to n variables or 'unknowns'. A simple example might be to find x and y given:

$$2x + 3y - 3 = 0$$
$$3x + 2y - 5 = 0 \qquad (10.1)$$

An example that appears more complex might be to find the three T values such that

$$\mathrm{T}_1 = mg$$
$$T_2 \cos \alpha - T_3 \cos \beta = 0$$
$$T_2 \sin \alpha - T_3 \sin \beta - T_1 = 0 \qquad (10.2)$$

where $\alpha = 60°$, $\beta = 30°$, $m = 10$ and $g = 9.81$.

The use of matrix algebra to solve systems of equation is shown in later exercises.

Of course, we would solve these problems using the techniques we learnt in school. Setting up a worksheet to solve a set of simultaneous equations with only two or three unknowns would be an inefficient use of time. It would be worth setting up a worksheet if we had a large number of similar sets of equations to solve. Furthermore, the techniques learnt with only three unknowns can be expanded to cases of more variables. So let us look at how we might use Solver to find the values of x and y in the first example.

Summing the absolute values would also work but there are mathematical arguments that make it better to sum squares.

Our use of Solver so far has involved single equations. We have had one *Set Cell* and one *Changing Cell*. The latter was changed to give the former a near-zero value. Can we apply this method to the two equations and ask Solver to vary both x and y? The answer is a partial yes. We can have more than one *Changing Cell* but there can be only one *Set Target Cell*. If we have a third cell that sums the values of the two functions, we could get Solver to use this for its *Set Cell*. That should seem to work; if each function evaluates to zero then their sum should be zero. However, there is a problem. What if one cell evaluates to –4 and the other to +4? These will sum to zero but we will not get the correct values for the variables. Rather than sum the values of the two functions, we will sum the squares of the values to avoid this 'cancellation' problem.

Exercise 7: A Simple Simultaneous Equations Problem

We now create a worksheet to solve the simultaneous equations in Equation 10.1.

(a) Open CHAP10.XLS and on Sheet7 set up a worksheet similar to that in Figure 10.16. The cells with formulas are:

C4: =2*B4 + 3*B5 – 3 first equation in Equation 10.1
C5: =3*B4 + 2*B5 – 5 second equation in Equation 10.1
D4: =C4^2
D5: =C5^2
D6: =D4+D5

	A	B	C	D
1	System of Linear Equations			
2				
3	Variables		Equations	Squares
4	x	0	-3	9
5	y	0	-5	25
6			SUM =	34

Figure 10.16

(b) Make D6 the active cell and start Solver. The *Set Target Cell* should contain D6. Set *Equal to Value* 0 and in the *By Changing Cells* box, either type B4:B5 or use the mouse to select them. Alternatively, press the Guess button, and Solver will correctly set the output cells to $B$4:$B$5.

Solver has improved with each new version of Excel. To solve this problem in Excel 95 one had to run Solver twice!

(c) Press Solver's *Solve* button. Solver reports it has found a solution with 6.8×1^{-13} as the value in D6. This is reasonably close to zero. The reported values for *x* and *y* are 1.79999988 and –0.20000011, respectively. If you replace these with 1.8 and –0.2, the equations each evaluate to zero. So Solver is useful but not perfect.

If you open the Solver Options box and select Central Derivative, Solver gets a result that is even closer to the exact solutions when run with the initial zero values for the variables.

Exercise 8: An Improved Simultaneous Equations Solver

With Exercise 7 behind us we can tackle a more complex problem. Our task is to construct a worksheet that will help us solve any system of linear equations with three unknowns. Let the system have the general form:

$$a_1x + b_1y + c_1z - d_1 = 0$$
$$a_2x + b_2y + c_2z - d_2 = 0$$
$$a_3x + b_3y + c_3z - d_3 = 0$$

We shall use three named cells for the variables *x*, *y* and *z*, and 12 named cells for the coefficients. Our worksheet will show cells with text such as 'a1', 'a2', etc., but the names will, of course, be 'a1_', 'a2_', etc. We cannot use 'c1', etc., as names so we will use 'c_1', etc., both as text and as names of cells.

As a concrete example upon which to design the worksheet, we will solve the system of equations in Equation 10.3.

$$\begin{aligned} 2x + 4y + 5z - 33 &= 0 \\ 6x + 6y + 7z - 70 &= 0 \\ 3x - 6y + 4z + 71 &= 0 \end{aligned} \tag{10.3}$$

(a) On Sheet8 of CHAP10.XLS construct a worksheet as shown in Figure 10.17 temporarily ignoring K4:K7. The values of 1 for the variables are arbitrary starting values.

(b) We now need to use Insert|Name|Create five times; once for A4:B6, then with C4:D6, etc.

	A	B	C	D	E	F	G	H	I	J	K
1	Simultaneous Equations Solver										
2											
3	Variables					Coefficients					Formulas
4	x	1	a1	2	b1	4	c_1	5	d1	-33	-22
5	y	1	a2	6	b2	6	c_2	7	d2	-70	-51
6	z	1	a3	3	b3	-6	c_3	4	d3	71	72
7								Sum of Squares			8269

Figure 10.17

(c) Enter the formulas for the three equations in the set:
K4: =a1_*x + b1_*y + c_1*z + d1_
K5: =a2_*x + b2_*y + c_2*z + d2_
K6: =a3_*x + b3_*y + c_3*z + d3_

(d) As before, we need a single cell as the target for Solver. Rather than computing the sum of the absolute values of the three functions, we will sum their squares with a single formula. In K7 type the formula SUMSQ(K4:K6) to achieve this.

(e) We are ready to use Solver. The *Target* cell should be K7 – the cell that sums the squares of the functions. Make sure to use *Equal to Value* of 0, and set the *By Changing Cells* to B4:B6.

Changing the option to use Central Derivatives does involve more computational steps but it improves the accuracy of the result with problems such as these.

Open the Option dialog, locate the Derivative area and select Central. Return to the main Solver dialog by clicking OK.

(f) Now click the Solve button. The results are $x = 6.0$, $y = 11.5$ and $z = -5.0$. The first equation evaluates to -1.6×10^{-13}, the other two to zero.

(g) Save the workbook.

Exercise 9: Non-linear Simultaneous Equations Solver

The simultaneous equations we have solved so far were all linear. These could have been solved using simple algebraic techniques. Non-linear systems of equations are far more difficult to solve with paper and pencil, requiring a knowledge of calculus. Let's see if Solver is capable of coming to our aid. We will solve the non-linear simultaneous equations:

$$x^2 + 2y^2 - 22.0 = 0$$
$$-2x^2 + xy - 3y + 11.0 = 0$$

(a) Figure 10.18 shows the required worksheet with the cell contents displayed as formulas. Set this up on Sheet9 of CHAP10.XLS. Note that cells in B3:B4 have the names shown to their left. The required formulas are:
D3: =x^2 + 2*y^2 – 22
D4: =–2*x^2 + x*y – 3*y + 11
D5: =SUMSQ(D3:D4)

	A	B	C	D
1	System of Non-linear Equations			
2				
3	x	1	f(x)	-19
4	y	1	g(x)	7
5			SumSq	410

Figure 10.18

(b) Call up Solver. We need to make D5 the target cell and B3:B4 the cells to be varied to equate the target cell to 0. Again it will be useful, but not essential, to use central derivatives.

With starting values of 1 for both variables Solver suggests an approximate solution with $x = 1.99994$ and $y = 3.00004$. The function evaluate to values somewhat larger than 0. Experimentation shows $x = 2$ and $y = 4$ is an exact solution. Of course, since the equations contain x^2 and y^2, multiple solutions are possible. Starting with 0 for each variable, Solver reported $x = -0.2763$ and $y = 3.3109$ as a solution.

(c) Save the workbook.

Concepts C: Matrix Algebra

The Microsoft Excel matrix functions are:

MDETERM(A)	Returns the matrix determinant of an array.
MINVERSE(A)	Returns the inverse of the matrix of an array.
MMULT(A,B)	Returns the matrix product.
TRANSPOSE(A)	Returns the transpose of an array. The first row of the input array becomes the first column of the output array, etc.

Some points to watch for when using matrix functions:

(i) For all but TRANSPOSE(), every cell in the array must contain a numeric value otherwise the #VALUE! error results.

(ii) The #VALUE! error results when an illegal matrix operation is attempted.

(iii) Because the first two functions use an accuracy of 16 digits, small numeric errors may occur. For example, a singular array may return a result that differs from zero by 1E–16.

(iv) Except for MDTERM(), these are array functions and must be completed with Ctrl+Shift+Enter.

Before we use these functions, a brief review of matrix algebra may help.

If when using a matrix formula you get one value rather than many, then you have forgotten either (i) to select a range before entering the formula or (ii) to complete the entry using Ctrl+Shift+Enter.

1. A matrix with m rows and n columns is said to be of order $m \times n$. When n and m are equal, the matrix is said to be square.

2. Matrix A is said to be identical to matrix B when every element in A equals the corresponding element in B, i.e. $a_{ij} = b_{ij}$. Clearly the two matrices must have the same order.

3. If A and B are of the same order, then $A + B = C$, and C has the same order as A and B. The elements in C are found using $c_{ij} = a_{ij} + b_{ij}$. Matrix subtraction is defined in a manner similar to addition.

4. If A is a matrix of order $m \times n$ and B is a matrix of order $n \times p$ (i.e. the number of columns in A equals the number of rows in B), then the *matrix product* of A and B, AB, is defined and is a matrix of order $m \times p$. If $AB = C$, then

$$c_{ij} = \sum_{k=1}^{n} a_{ik} b_{kj}$$

5. Matrix multiplication is not commutative; AB does not necessarily equal BA.

6. Matrix division is undefined.

7. An *identity matrix*, I, is a square matrix in which the elements on the main diagonal have values of 1 and all others have values of zero.

8. Let A be a square matrix. Then the *inverse* of A, represented by A^{-1}, is defined as the matrix which when multiplied by A yields an identity matrix. Thus, we see $AA^{-1} = I$. If B is the inverse of A, then A is the inverse of B and $BA = I = AB$.

9. A matrix which has no inverse is said to be *singular.*

10. When a matrix A is multiplied by an identity matrix, the result is A; $IA = A$.

Exercise 10: Some Matrix Operations

In this exercise we demonstrate some of the topics reviewed above. We will find the sum of two matrices, find a matrix inverse, and show that $AA^{-1} = I$.

(a) Open the workbook CHAP10.XLS and move to Sheet10. Enter the text and values shown in A1:G3 of Figure 10.19. Use the *Merge and Center* tool to centre the text over two columns.

(b) Enter the values in A4:E5.

(c) We now find the elements of C such that $A + B = C$. In G4 enter the formula =A4+D4. Copy this to G4:H5.

	A	B	C	D	E	F	G	H
1	Matrix operations							
2								
3	Matrix A			Matrix B			Matrix C = A+B	
4	2	3		7	8		9	11
5	3	4		9	10		12	14
6								
7	Matrix D			Matrix D^{-1}			Matrix $D*D^{-1}$	
8	1	-2		0.4	0.2		1	-2.8E-17
9	3	4		-0.3	0.1		0	1
10								
11	Determinant							
12	10	Non singular						

Figure 10.19

(d) Enter the text row 7 and centre each across two columns. To get superscripts, select the text (such as –1) and use the command Format|Cells and put a check mark in the *Superscript* box on the Font dialog box.

(e) Enter the values in A8:B9. These are the elements of matrix *D*.

(f) To compute the inverse of *D*, select the range D8:E9. Enter the formula =MINVERSE(A9:B9) and press Ctrl+Shift+Enter.

(g) To show that $DD^{-1} = I$, select G8:H9. Enter the formula =MMULT(A8:B9, D8:E9) and press Ctrl+Shift+Enter. You may wish to use the Formula Palette: type =MMULT, press Ctrl+A. Do not click the OK button on the Formula Palette, use Ctrl+Shift+Enter.

Note that the main diagonal elements have values of 1. We expected the off-diagonal elements to have values of zero but in H8 we have –2.8E–17. Recalling restriction (iv) above (Microsoft Excel uses a precision 1E–16 in computing matrix functions) we shall accept this value as zero. If you replace the values in the first row by –1 and 2, and in the second row by 2 and 5, the off-diagonal elements compute to identically zero.

(h) The formula =MDETERM(A8:B9) in A12 computes the determinant of the square matrix *D*. No inverse matrix exists for a matrix with a zero determinant. The formula in B12 is =IF(A12=0,"Singular","Non-singular").

(i) Save the workbook.

Exercise 11: Solving Systems of Linear Equations

A system of linear equations may be represented in matrix form. Thus the system of two equations:

$$x - 2y = -1$$
$$3x + 4y = 17$$

may be represented by:

$$\begin{bmatrix} 1 & -2 \\ 3 & 4 \end{bmatrix} \begin{bmatrix} x \\ y \end{bmatrix} = \begin{bmatrix} -1 \\ 17 \end{bmatrix} \qquad \text{Equation M}$$

Performing the matrix multiplication in Equation M, we get:

$$\begin{bmatrix} x - 2y \\ 3x + 4y \end{bmatrix} = \begin{bmatrix} -1 \\ 17 \end{bmatrix}$$

When matrix A and matrix B are equal, the corresponding elements are equal. So it follows that $x + 2y = 14$ and $2x - y = 5$; these are the equations with which we started, thereby justifying the statement that we may represent a system of linear equations in matrix form.

In the following A represents the matrix of the coefficients, X the matrix of the variables and C the matrix of the constants.

Let us write Equation M in the form:	AX	=	C
Multiply both sides by the inverse of A giving	$A^{-1}AX$	=	$A^{-1}C$
But since $A^{-1}A = I$, this becomes	IX	=	$A^{-1}C$
We know that $IX = X$, therefore	X	=	$A^{-1}C$

From this we see that the value of the X matrix may be obtained by computing $A^{-1}C$. In Exercise 10 we found the inverse of this matrix, so we may write:

$$\begin{bmatrix} x \\ y \end{bmatrix} = \begin{bmatrix} 0.4 & 0.2 \\ -0.3 & 0.1 \end{bmatrix} \begin{bmatrix} -1 \\ 17 \end{bmatrix}$$

Performing the multiplication gives

$$\begin{bmatrix} x \\ y \end{bmatrix} = \begin{bmatrix} 3 \\ 2 \end{bmatrix}$$

From which we see that $x = 3$ and $y = 2$.

This may have left you less than impressed; you could have solved the two simultaneous equations in your head. But the method may be applied to more challenging problems. In this exercise we solve:

$$\begin{aligned} 2x + 3y - 2z &= 15 \\ 3x - 2y + 2z &= -2 \\ 4x - y + 3z &= 2 \end{aligned}$$

The completed worksheet will resemble that in Figure 10.20.

(a) Open workbook CHAP10.XLS and move to Sheet11. Enter the text in rows 1, 2, 3, 9 and 14.

(b) In A5:C7 enter the coefficients of the equations, and in D5:D7 enter the constants.

If typing (C) results in the copyright symbol ©, use Tools|AutoCorrect to remove this from the list of automatic corrections.

(c) The next step is to compute the inverse (A^{-1}) of the matrix of coefficients. Select the range A10:C12, enter the formula =MINVERSE(A5:C7) and press Ctrl+Shift+Enter. The cells shown in the figure were formatted to display five places.

(d) The final step to find the solutions is to compute $A^{-1}C$. Select D10:D12, enter the formula =MMULT(A10:C12, D5:D7) and press Ctrl+Shift+Enter.

The solutions have now been found. We may wish to check that these agree with the system of equations.

(e) Use Insert|Name to name the cells D5:D7 as *x*, *y* and *z*, respectively.

(f) The formulas in row 15 are:
A15: =A5*x
B15: =B5*y
C15: =C5*z
D15: =SUM(A15:C15) Use AutoSum to make this.

These formulas are copied down to row 17. The values in D15:D17 agree with those in D5:D7, thus confirming that we have solved the system of equations.

	A	B	C	D	E
1	Using Matrix Functions to Solve				
2	a System of Linear Equations				
3					
4	Matrix of Coefficients (A)			Matrix of Constants (C)	
5	2	3	-2	15	
6	3	-2	2	-2	
7	4	-1	3	2	
8					
9	Inverse A^{-1}			$A^{-1}C = X$	
10	0.19048	0.33333	-0.09524	2	x
11	0.04762	-0.66667	0.47619	3	y
12	-0.23810	-0.66667	0.61905	-1	z
13					
14	Reconstructed equations				
15	4	9	2	15	
16	6	-6	-2	-2	
17	8	-3	-3	2	

Figure 10.20

We now have a worksheet that may be used to find the solutions of any set of linear equations in three unknowns, providing the equations are independent and consistent. If the equations are not independent (for example, the third equation is twice the first plus the second) then the A matrix will be singular – it will have no inverse. In this case Microsoft Excel will return the #NUM! error value when MINVERSE is used with a singular matrix. The MDETERM function may also be used to test the A matrix; MDETERM(array) returns a zero value for a singular matrix. If the determinant of A is very small it may be difficult to solve the set of equations.

Concepts D: Curve Fitting

In Chapter 7 we learned how to fit data to a variety of functions using the trendline chart feature and worksheet functions such as SLOPE, LINEST and LOGEST. We also saw how to linearize functions to enable these methods to be applicable. However, certain experiments may need alternative methods.

	A	B	C	D
1	Least squares fit using Solver			
2	m	1		
3	b	0		
4	resSq	59.67		
5	x	y	y'	$resid^2$
6	0.2	0.1	0.200	0.010
7	0.3	0.3	0.300	0.000
8	0.3	0.6	0.300	0.090
9	0.4	0.3	0.400	0.010
10	0.5	0.2	0.500	0.090
11	0.6	0.1	0.600	0.250
12	10.1	9.2	10.100	0.810
13	11.1	10.2	11.100	0.810
14	11.6	10.8	11.600	0.640
15	118.2	118.1	118.200	0.010
16	118.3	117.6	118.300	0.490
17	120.2	119.6	120.200	0.360
18	226.5	228.1	226.500	2.560
19	228.1	228.3	228.100	0.040
20	229.2	228.9	229.200	0.090
21	337.4	338.8	337.400	1.960
22	338	339.3	338.000	1.690
23	339.1	339.3	339.100	0.040
24	447.5	448.9	447.500	1.960
25	448.6	449.1	448.600	0.250
26	448.9	449.2	448.900	0.090
27	556	557.7	556.000	2.890
28	556.8	557.6	556.800	0.640
29	558.2	559.2	558.200	1.000
30	666.3	668.5	666.300	4.840
31	666.9	668.8	666.900	3.610
32	669.1	668.4	669.100	0.490
33	775.5	778.1	775.500	6.760
34	777	778.9	777.000	3.610
35	779	778.9	779.000	0.010
36	884.6	888	884.600	11.560
37	887.2	888	887.200	0.640
38	887.6	888.8	887.600	1.440
39	995.8	998	995.800	4.840
40	996.3	998.5	996.300	4.840
41	999	998.5	999.000	0.250

(A) Linear data

	A	B	C
1	Gaussian Fit		
2			
3	h	1600	
4	mu	0.255	
5	sig	0.005	
6	base	0	
7	resSq	1620774	
8			
9	x	y	y'
10	0.239	25	0.06
11	0.240	24	0.20
12	0.241	39	0.63
13	0.242	49	1.85
14	0.243	56	5.04
15	0.244	84	12.65
16	0.245	66	29.31
17	0.246	97	62.66
18	0.247	158	123.69
19	0.248	244	225.37
20	0.249	353	379.08
21	0.250	444	588.61
22	0.251	773	843.67
23	0.252	1196	1116.28
24	0.253	1677	1363.43
25	0.254	1654	1537.26
26	0.255	1341	1600.00
27	0.256	1173	1537.26
28	0.257	933	1363.43
29	0.258	550	1116.28
30	0.259	220	843.67
31	0.260	101	588.61
32	0.261	97	379.08
33	0.262	39	225.37
34	0.263	26	123.69
35	0.264	11	62.66
36	0.265	16	29.31
37	0.266	10	12.65
38	0.267	13	5.04
39	0.268	8	1.85
40	0.269	5	0.63

(B) Non-linear data

Figure 10.21

When we perform a linear least squares fit, we establish which values of m and b in the equation $y = mx + b$ best fit the experimental data. By 'best fit' we mean gives the smallest value to the quantity $\Sigma(y_i - y'_i)^2$ where y_i and y'_i are the experimental and the predicted values, respectively. Suppose our data was to be fitted to a non-linear function having the parameters a, b and c. A least squares fit would similarly find which values of parameters minimized the sum of the squares of the difference between the experimental and the predicted values. This sounds very much like something Solver could do.

We will test this hypothesis in Exercise 12 by fitting the data in Figure 10.21A to a linear function and comparing Solver's results with those obtained with the earlier methods. We will then use Solver in Exercise 13 to fit the non-linear data in Figure 10.21B.

In Figure 10.21A we have experimental *x,y* data in A6:B41. Using the slope m and intercept b values in B2:B3, we compute the fitted data y' in C6:C41 using $y' = mx + b$. Of course, our guess of 1 and 0, respectively, are wildly wrong. In D6:D41 we compute the individual residuals squared and these are summed in B4. Solver's task will be to vary the slope and intercept to minimize this sum.

Figure 10.21B is similar except that we have four parameters (*h*, *mu*, *sig* and *base*). Rather than compute the individual residuals squared and sum them, we use an Excel function to get the required $\Sigma(y_i - y'_i)^2$ in B7.

Exercise 12: A Linear Curve Fit

Our purpose is to compare the results obtained by fitting the data to a linear function using Solver with those from the SLOPE and INTERCEPT functions.

(a) On Sheet12 of CHAP10.XLS start a worksheet by entering the text shown in the cells of Figure 10.20. Name the cells B2:B3 with the names in column A. Enter the values 1 and 0 in B2 and B3, respectively. These are our starting values for the fit.

(b) Enter the *x* and *y* values in A6:A41 and B6:B41, respectively.

(c) The formula in C6 used to compute the predicted value is =m*A6+b. In D6 we compute the square of the residual using =(B6-C6)^2. These are copied down to row 41.

(d) In B4 enter =SUM(D6:D41) to compute the sum of the squares of the residuals – this quantity is sometimes referred to as SSR.

(e) Invoke Solver with B4 as the target cell and B2:B3 as the cells to be changed. Since there is no way that Solver can adjust the slope and intercept to lower the SSR value to zero there is no point in asking Solver to make the target cell zero. Instead we will ask it to minimize the target cell.

(f) It is left as an exercise for the reader to use the SLOPE and INTERCEPT (or LINEST) functions to get values to compare with Solver's *m* and *b* values. The comparison is made in Figure 10.24. While the values are so similar as to be identical for most practical purposes, you may wonder which set of values is better. The data used in this exercise came from the Norris database provided by the National Institute of Science and Technology (NIST) on its web site (http://www.nist.gov). This also provides the certified values for the fit which are shown in Figure 10.22. The worksheet function results are in slightly better agreement with the NIST values.

	Slope	Intercept
Solver	1.00211680038825	-0.262323072920204
Formulas	1.00211681802045	-0.262323073774041
NIST	1.00211681802045	-0.262323073774029

Figure 10.22

Exercise 13: A Gaussian Fit with Solver

The function used to fit experimental data to a Gaussian curve is:

$$y_i = h\exp\left(-\left(\frac{x_i-\mu}{\sigma}\right)^2\right) - b$$

where:

y_i = the predicted value
h = the peak height above the baseline
x_i = the value of the independent variable
μ = the position of the maximum
σ = the standard deviation and
b = the baseline offset

We shall fit the 31 data points in Figure 10.23 to a Gaussian curve by varying *h*, μ, σ and *b*. Clearly, 31 data points is too few for a good fit when four parameters are involved, but it will show the procedure without overtaxing the reader with too much data entry.

(a) On Sheet13 of CHAP10.XLS start a worksheet similar to that in Figure 10.23. Begin by entering the text in rows 1 to 9, and naming the cells B3:B6 with the text in column A. Temporarily ignore the values in B3:B7.

(b) Enter the x and y values in rows 10 to 40 and construct a chart of the data. This will resemble the markers in the left-hand chart in Figure 10.25.

The starting values (B3:B6) are more critical than before. We are working with a more complex function and have more variable cells. If we start too far from the answer Solver will reply 'You can't get there from here'. Looking at the chart, a value of 1600 for h seems reasonable. The curve has its peak more or less midway between 2.5 and 2.66, so we will start with 0.255 in B4. Recalling that three standard deviations on each side of the mean encompasses most of a Gaussian curve, we compute a starting value for the standard deviation (B5) as =(0.27 – 0.24)/6 = 0.005.

The tails of the curve are not far from zero, so a starting value of 0 for b would be appropriate.

(c) With these values entered into B3:B6, enter this formula into C10 to compute the predicted value: =h*EXP(-(((A10-mu)/sig)^2)) + base. You may feel there is an extra set of parentheses but this is not so for we must allow for the fact that the negation operator has the highest priority. Copy the formula down to row 40.

(d) Select C10:C40 and use the Copy tool. Click on the chart and use Paste Special to add a new data series. Your chart should now resemble the left-hand chart in Figure 10.23. Note how our analysis has generated a good starting fit to the experimental data.

(e) Rather than adding another column to compute the squares of the residuals and adding these to get SSR, we will use the SUMXMY2 (Sum X Minus Y power of 2) function provided by Excel. This sums the squares of the differences of two ranges – just what we need. For our problem the two ranges are the y and y' ranges. In B7 enter the formula =SUMXMY2(B10:B40,C10:C40).

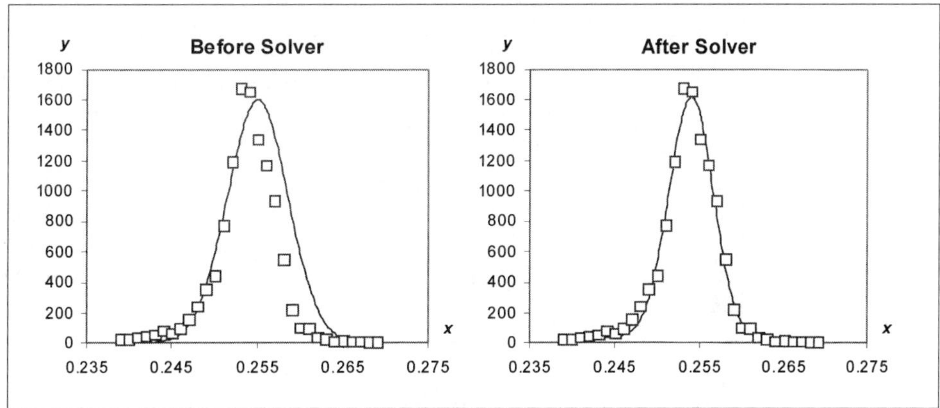

Figure 10.23

(f) Use Solver to complete the task. The target cell is B7 which we wish to minimize by changing B3:B6. The resulting values are shown below.

h	μ	σ	base	resSqr
1580.69	0.253959	0.003654	40.11852	114867.2

If you wish to experiment with larger, more realistic data sets and more complex problems (i.e. two overlapping Gaussian curves), the NIST web site provides some fine examples. Furthermore, it provides the values you should get for the parameters. Visit Chapter 14 if you wish to experiment with fitting a histogram to a normal curve.

Matrix Diagonal

Although this chapter is about solving equations, we did touched on matrix algebra so it seems appropriate to demonstrate how to sum a matrix diagonal.

In Figure 10.24 the range A1:C3 is named *Amat* while C8:E10 is named *Bmat*. The position of *Amat* makes for a simple array formula in C5 since the first element of the matrix is in row 1, column 1. When the matrix is positioned more arbitrarily, one must adjust the formula as shown in C12. The need to offset the ROW and COLUMN values should immediately alert us to the fact that moving the matrix will invalidate the array formula. We may avoid this worry if we are prepared to use three cells to perform the computation – see the lower part of the figure. When you have read Chapter 8, you may wish to code a user-defined function in VBA to perform this calculation.

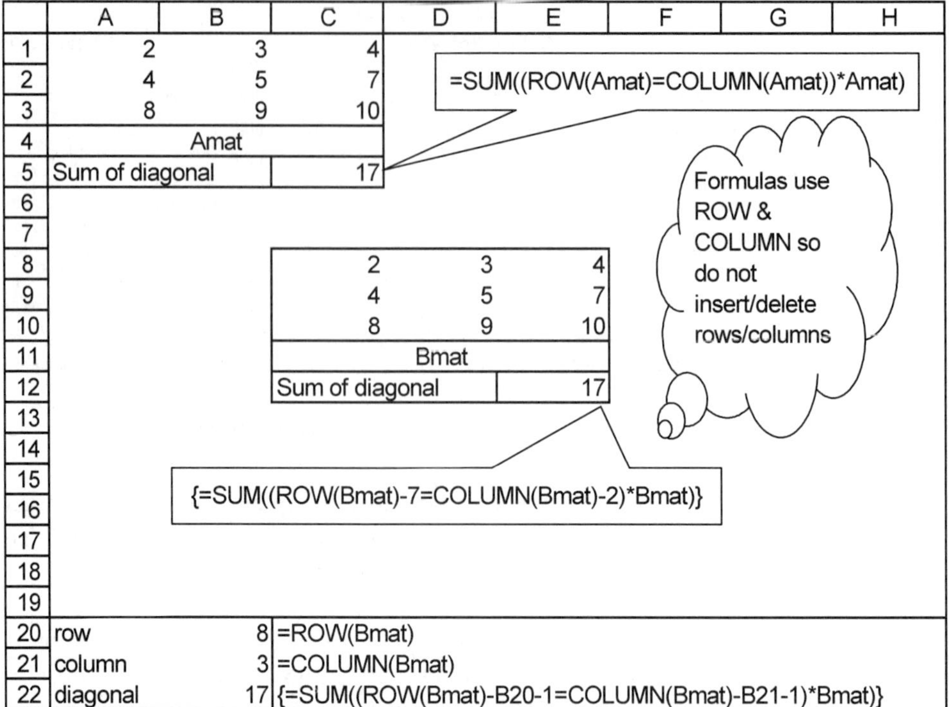

A B C D E F G H
1 2 3 4
2 4 5 7
3 8 9 10
4 Amat
5 Sum of diagonal 17
=SUM((ROW(Amat)=COLUMN(Amat))*Amat)
Formulas use ROW & COLUMN so do not insert/delete rows/columns
2 3 4
4 5 7
8 9 10
Bmat
Sum of diagonal 17
{=SUM((ROW(Bmat)-7=COLUMN(Bmat)-2)*Bmat)}
20 row 8 =ROW(Bmat)
21 column 3 =COLUMN(Bmat)
22 diagonal 17 {=SUM((ROW(Bmat)-B20-1=COLUMN(Bmat)-B21-1)*Bmat)}

Figure 10.24

Problems

1. The equations below may look like 'made-up' problems. Actually, many of them come from real physics or engineering problems. Use Goal Seek to find the solutions to the following:

   (a) $x^2 + 1 = 2e^{-x}$

   (b) $z\sin(z) - \cos(z) = 0$

   (c) $\sqrt{\dfrac{\sin(x)}{x}} = \dfrac{1}{2}$

   (d) $5\exp(-5t)\sin(\pi/4) + 5\sin(5t - \pi/4) = 0$

2. Use Solver to find solutions to the equations in (1) above. Compare the results with those obtained with Goal Seek. When solving (d), use constraints to specify $0.5 < t < 1.4$.

3.* Use Solver to find the answer to Problem 1 in Chapter 9.

4. A chemistry problem dealing with equilibrium constants requires a positive value of x such that:

   $$\frac{(0.22 + 2x)^2}{(1.00 - x)(0.70 - x)} = 0.10$$

   Note: you could expand the equation to form a quadratic or you could put the right-hand expression in the input cell for Solver and request Solver to vary the cell containing the x value to give the input cell a value of 0.10.

5. Expand the worksheet constructed in Exercise 8 such that it will solve a system of simultaneous equations with four unknowns. Use the worksheet to solve:

   $$\begin{aligned} x - y - z - 2w &= -1 \\ x - y + z + 4w &= -6 \\ 3x + y - z + 2w &= -4 \\ 5x + y - 3z + w &= -9 \end{aligned}$$

6. Write a VBA function to implement the bisection method shown in Exercise 1.

11
Numerical Integration

Concepts

Numerical integration is used to evaluate a definite integral when there is no closed-form expression for the integral or when the explicit function is not known and the data is available in tabular form only. Numerical integration (or *quadrature*) consists of methods to find the approximate area under the graph of the function $f(x)$ between two x values.

The simplest of these uses the *trapezoid rule*. If we divide the area under the curve into a sufficiently large number of parts, as shown in Figure 11.1, then the area under the curve (the approximate integral) is given by:

$$I = \int_a^b f(x)dx \approx \sum_{i=1}^{n} A_i \tag{11.1}$$

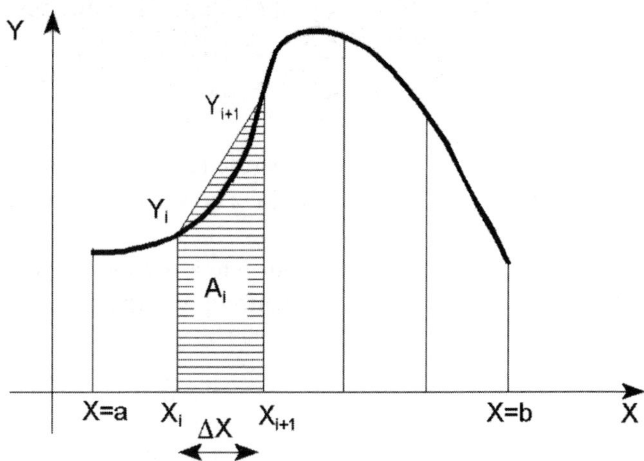

Figure 11.1

We approximate the representative strip to a trapezoid. For a clearer drawing, only five strips are used. Obviously more, smaller, strips are needed for a good approximation. Let there be n strips and hence $n+1$ data points. The area of a typical strip is:

$$A_i = \Delta x \frac{y_i + y_{i+1}}{2} \tag{11.2}$$

Combining the two equations we see that:

$$I \approx \Delta x \frac{y_1 + y_2}{2} + \Delta x \frac{y_2 + y_3}{2} + \cdots + \Delta x \frac{y_n + y_{n+1}}{2} \tag{11.3}$$

Giving:

$$I \approx \frac{\Delta x}{2}(y_1 + 2y_2 + 2y_3 + 2y_4 + \cdots + 2y_n + y_{n+1}) \tag{11.4}$$

or

$$I \approx \frac{\Delta x}{2}\left(y_1 + 2\sum_{i=2}^{n} y_i + y_{n+1} \right) \tag{11.5}$$

Equation 11.5 is called the *trapezoid rule.* We use the trapezoid approximation in the first exercise to evaluate an integral.

A better approximation to the integral is obtained by taking two adjacent strips and joining the three points on the curve with a parabola. This gives Equation 11.6, called the *Simpson one-third rule* for approximating the area under a curve.

$$I \approx \frac{1}{3} \sum_{i=1,3,5}^{n-2} (y_i + 4y_{i+1} + y_{i+2}) \Delta x \tag{11.6}$$

This rule requires that there be an even number of equally spaced strips. The second exercise uses this approximation.

Approximating the curve through *four* adjacent points to a cubic equation gives the *Simpson ⅜ rule* shown in Equation 11.7.

$$I \approx \frac{3}{8} \sum_{i=1,4,7}^{n-3} (y_i + 3y_{i+1} + 3y_{i+2} + y_{i+3}) \Delta x \tag{11.7}$$

Surprisingly, the ⅜ rule is often less accurate than the ⅓ rule. However, unlike the ⅓ rule, it does not require an even number of strips. This is an advantage when the data is available only in tabular form (the explicit function being unknown) and there is an even number of data points – an odd number of strips.

Accuracy

If the interval *a* to *b* is divided into successively more strips then, in principle, the accuracy should increase. This is not so in practice. When the number of strips becomes very large, the accumulated round-off error becomes significant.

In some of the exercises and problems, we evaluate definite integrals with known values. There are two reasons for doing this: we can compare our approximations with the known values to check the accuracy, and it will give us confidence to attack integrals with unknown values.

Exercise 1: The Trapezoid Rule

Using the trapezoid approximation, we evaluate the integral

$$\int_0^{\pi} x\sin(x)dx$$

and show that the approximation yields a result close to the exact result of π. Looking at Figure 11.1 and Equation 11.3, we see that when five strips are used a total of six $f(x)$ or y-values are needed. In general, when n strips are used then $n+1$ values of y are required. We start this exercise with 10 strips so we need 11 values of y. For this exercise we will use the trapezoid rule in the form of Equation 11.3. When you have completed this exercise your worksheet should resemble that in Figure 11.2.

(a) Open a new workbook and enter the text in A1:A6. Enter the following:

B3:	0	The lower limit of the integration
B4:	=PI()	The upper limit of the integration
B5:	10	The number of strips
B6:	=(upper-lower)/B5	

The last formula computes the value of Δx which we use to find the 11 x values: *lower*, *lower* + Δx, *lower* + $2\Delta x$, etc.

(b) Select A3:B6 and use Insert|Name|Create to name the cells in B3:B6.

(c) Enter the text in row 8.

(d) Enter the following formulas:

A9:	=lower	The value of x_1.
A10:	=A9+delta	The value of x_2

Copy these down to row 19 to give the 11 x values.

(e) In B9 enter the formula =A9*SIN(A9) to compute the value of y_1. Copy this formula down to B19. This gives the 11 y values for the strips. Note how using the Equation 11.3 form of the trapezoid rule allowed us simply to copy B9 to the other cells.

	A	B	C
1	Trapezoid Rule		
2			
3	lower	0	
4	upper	3.141593	
5	n	10	
6	delta	0.314159	
7			
8	x	y	strip
9	0	0	0.015249
10	0.314159	0.097081	0.073261
11	0.628319	0.369316	0.177782
12	0.942478	0.762481	0.307501
13	1.256637	1.195133	0.434471
14	1.570796	1.570796	0.528337
15	1.884956	1.792699	0.56106
16	2.199115	1.779121	0.511512
17	2.513274	1.477265	0.369293
18	2.827433	0.873725	0.137244
19	3.141593	3.85E-16	
20		Approx	3.115711
21		Exact	3.141593
22		Error	-0.82%

Figure 11.2

(f) In C9 enter the formula =delta*(B9+B10)/2 to compute the area of the first strip. Copy this formula down to C18 to get the areas of the other nine strips. Be careful NOT to copy it to C19.

(g) Enter the text in B20:B22.

(h) Enter the following formulas:

C20: =SUM(C9:C18)	To sum the 10 trapezoid areas
C21: =PI()	The exact result for the integral
C22: =(C20-C21)/C21	

The last formula computes the error in the approximation.

(i) Save the workbook as CHAP11.XLS.

If all has gone well you will see that the trapezoid rule approximates the integral to 3.115711, which is 0.82% low compared to the exact value. You may wish to modify the worksheet to use 20 strips and note how the error is reduced to −0.21%.

Exercise 2: Simpson's ⅓ Rule

This exercise uses Simpson's ⅓ rule to find the approximate value of

$$\int_0^2 \exp(x^2)dx$$

There is no analytical method for this integral but the published value for the interval [0,2] is 16.452627. For improved accuracy we will use 20 strips; which means we need 21 *x*,*y* pairs. When completed, your worksheet should resemble Figure 11.3.

	A	B	C	D
1	Simpson's ⅓ Rule			
2				
3	lower	0		
4	upper	2		
5	n	20		
6	delta	0.1		
7				
8		x	y	strip
9	1	0	1	6.081011
10	2	0.1	1.01005	
11	3	0.2	1.040811	6.591019
12	4	0.3	1.094174	
13	5	0.4	1.173511	7.742942
14	6	0.5	1.284025	
15	7	0.6	1.433329	9.859075
16	8	0.7	1.632316	
25	17	1.6	12.93582	110.4428
26	18	1.7	17.99331	
27	19	1.8	25.53372	227.9961
28	20	1.9	36.96605	
29	21	2	54.59815	
30			Approx	16.45521

Figure 11.3

(a) Open the workbook CHAP11.XLS and on Sheet2 enter the text in A1:A6.

(b) Select A3:B6 and use the command Insert|Name|Create to name the cells in B3:B6.

(c) Enter the following values or formulas:

B3:	0	The lower limit of the integration
B4:	2	The upper limit of the integration
B5:	20	The number of strips
B6:	=(upper-lower)/n	

The formula in B6 computes the value of Δx which we use to find the 21 x -values: *lower*, *lower* $+\Delta x$, *lower* $+2\Delta x$, etc.

(d) Enter the text in row 8.

(e) The numbers in A9:A29 are not essential but may help you understand how the Simpson rule is implemented in the worksheet. Enter these using the Fill Series method you learnt in an earlier chapter.

(f) Enter the initial x and y values in B9 and C9:

B9:	=lower	The value of x_1
C9:	=EXP(B9^2)	The value of y_1

(g) Enter the second and subsequent values of x and y:

B10:	=B9+delta	The value of x_2
C10:	=EXP(B10^2)	The value of y_2

Copy B10:C10 down to row 29.

When the values of Δx values are constant Equation 11.6 becomes:

$$I = \frac{\Delta x}{3} \sum_{i=1,3,5}^{n-2} (y_i + 4y_{i+1} + y_{i+2}) \tag{11.8}$$

Thus we may compute each of the terms, find the summation and multiply the result by $\Delta x/3$ to approximate the integral.

The first term in the summation is $(y_1 + 4y_2 + y_3)$ so in D9 enter the formula =C9 + 4*C10 + C11. Check that your value agrees with that in Figure 11.3.

(h) We need to copy this formula to each alternate cell in the range D11:D27. The easiest way to do this is select D9:D10 and drag the fill handle down to D27.

(i) Enter the text in C30 and in D30 enter =SUM(D9:D29)*delta/3 to compute the integral.

(j) Enter appropriate text and formulas in C31:D32 to show the published value and compute the percentage error.

(k) Save the workbook CHAP11.XLS, as we will use it in the next exercise.

Exercise 3: Adding Flexibility

In the last exercise we developed a worksheet to evaluate a certain integral. Do we need to repeat all that work if we wish to evaluate another? We could, of course, edit the formula in C9 to reflect the new function to be integrated and copy this down to C29. Another way is to put the function in a module sheet and change the user-defined function each time we wish to evaluate a different integral.

(a) Open the workbook CHAP11.XLS. We wish to duplicate Sheet2. Hold down the Ctrl key and drag the Sheet1 tab to the right (you will see an icon of a sheet of paper overprinted with a + sign). Release the mouse button and tab labelled Sheet2 (2) will appear. Delete Sheet3 and rename Sheet2 (2) as Sheet3.

(b) Open the Visual Basic Editor with the command Tools|Macro|Visual Basic Editor or with the Alt+F11 shortcut. Insert a macro sheet of the CHAP11 project. Type the user-defined function shown below.

```
'Function to use with Simpson Rule worksheet
Function Simp(x)
        Simp = Exp(x ^ 2)
End Function
```

Remember, whenever you edit a user-defined function, you must recalculate the worksheet by pressing F9 before any changes in the function will take effect.

(c) Return to the worksheet containing the Simpson rule calculation. Change the formula in C9 to =SIMP(B9) and copy this down to C29. The values should stay the same as before (see Figure 11.3).

(d) Now we will make a quick change to evaluate

$$\int_{-1}^{1} \exp(-x^2)dx$$

The change made in the module will not be reflected in the worksheet until something happens to make Excel recalculate the worksheet. Pressing F9 will cause this, as will changing the values for the limits.

Change the third line in the module function to read Simp = Exp(–(x ^ 2)). Carefully note the position of the negation operator relative to the parentheses.

(e) Return to the Simpson worksheet. Change the values of *lower* and *upper* to −1 and 1, respectively. The value of the new function has been calculated. Remember, you may have to change the cell with the published value. For this function, in the interval −1 to 1, the result should be approximately 1.49.

(f) To make another test, change the user-defined function to Simp = x * Sin(x). Return to the Simpson worksheet and change the values of *lower* and *upper* to 0 and PI(), respectively. How does your result compare to that in Exercise 1?

(g) Save the workbook.

Exercise 4: Going Modular

In this exercise we implement the Simpson ⅓ rule as a user-defined function in a module. To check our function more easily we find the value of the same integral as in Exercise 1. In addition, we experiment with making the strip successively smaller and observe how the percentage error changes.

(a) Open the Visual Basic Editor with the Alt+F11 shortcut. On the module sheet for CHAP11 project, code the integrating function and the function to be integrated as in Figure 11.4. Do not type the line numbers; they are for discussion purposes only. The statements in the *Integral* function are examined at the end of the exercise.

(b) On Sheet4 of the CHAP11.XLS workbook enter the text and values shown in A1:A10 of Figure 11.5.

(c) Select A3:B5 and, using Insert|Name|Create, name the cells B3:B5.

(d) The formulas and values to be entered in B3:B10 are:

B3: 0 — The lower limit
B4: =PI() — The upper limit
B5: =PI() — The known value of the integral
B7: 10 — The number of strips
B8: =Integral(lower,upper,B7) — Value from Simpson's rule
B9: =B8−exact — Error calculation

B10: =B9/exact   Percentage error

Format B10 to show a percentage value with four decimal places.

(e) Check that your results in B8:B10 agree with Figure 11.5. If they do not, you may need to edit your module or your formulas.

```
'Simpson One-Third Rule Approximation
Function Integral(a, b, n)
   Integral = 0#
   delta = (b - a) / n
   x = a
   For i = 1 To n Step 2
      Term = y(x) + 4 * y(x + delta) + y(x + 2 * delta)
      Integral = Integral + Term
      x = x + 2 * delta
   Next i
   Integral = Integral * delta / 3
End Function

'The function to be integrated
Function y(x)
   y = x * Sin(x)
End Function
```

Figure 11.4

	A	B	C	D	E	F
1	Simpson's ⅓ Rule					
2	With User-defined Integral Function					
3	lower	0				
4	upper	3.141593				
5	exact	3.141593				
6						
7	n	10	100	1,000	10,000	100,000
8	Approx	3.141765	3.141593	3.141593	3.141593	3.141593
9	Error	0.000172	1.7E-08	1.68E-12	-1.2E-13	-3.5E-12
10	% Error	0.0055%	0.0000%	0.0000%	0.0000%	0.0000%

Figure 11.5

(f) Select B7:B10 and drag the handle to column F. Change your n values to match those in the spreadsheet. Note that large numbers may be entered with a comma to make them more readable. Do not be surprised if your worksheet takes some time to respond. Microsoft Excel has to do a large number of calculations. Save the workbook.

Note how the absolute error progressively decreases up to n = 10,000 and then increases for larger n values. Here we are seeing the accumulated round-off errors beginning to creep in.

Explanation of the Integral function:

Line	*Comment*
3	Type this as Integral = 0.0 to tell VBA that it is a real, not an integer, value. This initializes the value of the function to zero.
5	The lower limit of the integral is a.
6	We need to sum for odd values of i (i = 1, 3, 5, ..., n). The step 2 phrase achieves this. It is equivalent to copying the formula in D9 of Exercise 2 to alternate rows.
7	This finds the $(y_i + 4y_{i+1} + yi+_2)$ term for the area of a strip. We multiply by $\Delta x/3$ at the end of the calculation.
8	We keep a running total of the terms computed in line 7. We may read this as: New Integral value = Old Integral value + Term value.
9	This statement computes the x value for the next term.
10	When we have found all the partial areas, we multiply by $\Delta x/3$.

Exercise 5: Tabular Data

There are times when the data to be integrated comes from an experiment and the implicit function is unknown. Which of the three rules should be used to evaluate the integral?

Trapezoid	may be used with any data but is the least accurate.
Simpson ⅓	requires an even number of equally spaced strips.
Simpson ⅜	requires equally spaced x values, may be less accurate than the ⅓ rule.

Suppose we have 63 strips (i.e. 64 data pairs). We may use the ⅜ rule for the first three and the ⅓ rule for the remaining 60 strips.

We have seen that increasing the number of strips improves the accuracy of these approximations. With tabular data, this option is not available. While we cannot increase the number of data points since we do not know the function, we can decrease the number by doubling the width of each strip. Essentially this means we ignore every alternate data pair in the table. Obviously, our second value for the integral will be less accurate. This is where Romberg integration is useful. The Romberg integral is computed using:

$$I_R = I_h + \frac{I_h - I_{2h}}{2^n - 1}$$

where:

I_R = the improved value of the approximation
I_h = the value of the approximation with strips of width h
I_{2h} = the value of the approximation with strips of width $2h$
n = 2 for the trapezoid rule and 4 for Simpson's rules

In this exercise we use the trapezoid rule to find an approximation to some tabulated data. Clearly, we may use Romberg integration only when the strips are evenly spaced.

	A	B	C	D	E
1	Tabular Data				
2					
3	x	y	h=0.2	h=0.4	
4	1.8	6.050	6.0500	6.0500	
5	2.0	7.389	14.7780		
6	2.2	9.025	18.0500	18.0500	
7	2.4	11.023	22.0460		
8	2.6	13.464	26.9280	26.9280	
9	2.8	16.445	32.8900		
10	3.0	20.086	40.1720	40.1720	
11	3.2	24.533	49.0660		
12	3.4	29.964	29.9640	29.9640	Romberg
13		Integrals	23.9944	24.2328	23.91493

Figure 11.6

(a) On Sheet5 of the workbook CHAP11.XLS enter the text in A1:D3 as shown in Figure 11.6.

(b) Enter the values in A4:B12. This is the experimental data which we wish to integrate.

(c) We will use the trapezoid rule in the form of Equation 11.4 in this exercise. The required formulas for the C column are:
C4: =B4
C5: =2*B5 copy this down to C11
C12: =B12
These are the bracketed terms in Equation 11.4. The integral is completed by adding the entry:
C13: =0.2/2*SUM(C4:C12)

(d) In column D we will use strips of twice the width. Enter the formulas:

D4: =B4
D6: =2*B6 copy this to D8 and D10
D12: =B12
D13: =0.4/2*SUM(D4:D12)

(e) The Romberg integral is found with
E13: =C13 + (C13 - D13)/3.

Is the Romberg value a better approximation? The data is actually $y = \exp(x)$ with the values rounded to three decimal places. Therefore our result should be the integral

$$\int_{1.8}^{3.4} \exp(x)dx = \exp(3.4) - \exp(1.8) = 23.9144$$

In row 14, compute the percentage errors of the three values. Clearly, the Romberg value is the more accurate one.

(f) Save the workbook.

Improper Integrals

Provided $f(x)$ is continuous over the closed interval $[a, b]$, the integral

$$\int_a^b f(x)dx$$

is said to exist. Some integrals do not satisfy this criterion. An integral is said to be improper when: (i) a or b is infinity, or (ii) $f(x)$ becomes infinite at one or more values in the interval $[a, b]$. Some improper integrals do converge to a limit and simple substitution often converts an improper integral to a proper one.

If the origin is the singular point, try the substitution $x = t^a$ with a large enough. Thus, the substitution $x = t^2$ transforms

$$\int_0^1 \tan\left(\sqrt{x}\right)dx \quad \text{to} \quad \int_0^1 2t\tan\left(t\right)dt$$

When the interval is unbounded try these substitutions: $t = \arctan(x)$, $x = t^{-a}$, $\exp(-ax) = t$. For example, with the substitution $x = 1/t$

$$\int_0^\infty \frac{1}{x^3 + x^2 + 1}dx \quad \text{yields} \quad \int_0^1 \frac{t}{t^3 + t^2 + 1}dt$$

It is important to remember that the substitution formula $x = g(x)$ must be continuous over the limits of the integration. The substitution $x = 1/u$ would be acceptable when the bounds are 0 and +1 but not for −1 to +1.

Exercise 6: Gaussian Integration

The Gaussian two-point integration formula, as derived in most elementary numerical analysis textbooks, has the wonderful simplicity of:

$$\int_{-1}^{1} f(t)dt = f\left(-\tfrac{1}{\sqrt{3}}\right) + f\left(+\tfrac{1}{\sqrt{3}}\right) \quad (11.9)$$

The four-point formula is only slightly more formidable:

$$\int_{-1}^{1} f(t)dt = \tfrac{5}{9} f\left(-\tfrac{\sqrt{3}}{5}\right) + \tfrac{8}{9} f(0) + \tfrac{5}{9} f\left(+\tfrac{\sqrt{3}}{5}\right) \quad (11.10)$$

These formulas may be generalized to:

$$\int_{-1}^{1} f(t)dt = \sum_{i=1}^{n} w_i f(t_i) \quad (11.11)$$

The table below lists the values for the weights (w_i) and the points (t_i) for various numbers of points in the integration.

The degree of the polynomial function for which each integration formula is accurate is given by $2n-1$. Thus the three-point formula is accurate for polynomials up to degree 5. When one is unsure of the number of points to use, successively use 2, 3, … points until two results agree to the precision required.

The weights and point values in the table are for the limits of integration ±1. To use them with other limits (a to b) we make the substitution

$$x = \frac{(b-a)t + b + a}{2} \quad \text{so} \quad dx = \left(\frac{b-a}{2}\right)dt$$

giving

$$\int_{a}^{b} f(x)dt = \frac{b-a}{2} \int_{-1}^{1} f\left(\frac{(b-a)t + b + a}{2}\right)dt \quad (11.12)$$

n	$\pm t_i$	w_i
2	$\sqrt{1/3}=$ 0.57735 02691 89626	1
3	0	8/9 = 0.88888 88888 88889
	$\sqrt{3/5}=$ 0.77459 66692 41483	5/9 = 0.55555 55555 55555
4	0.33998 10435 84856	0.65214 51548 62546
	0.86113 63115 94052	0.34785 48451 37455
5	0	0.56888 88888 88889
	0.53846 93101 05727	0.47862 86704 99334
	0.90617 98459 38664	0.23692 68850 56189
6	0.23861 91860 83197	0.46791 39345 72691
	0.66120 93864 66264	0.36076 15730 48139
	0.93246 95142 03152	0.17132 44923 79170
8	0.18343 46424 95644	0.36268 37833 78363
	0.52553 24099 16329	0.31370 66458 77887
	0.79666 64774 13017	0.22238 10344 53966
	0.96028 98564 97439	0.10122 85362 90617
10	0.14887 43389 81631	0.29552 42247 14753
	0.43339 53941 29247	0.26926 67193 09997
	0.67940 95682 99032	0.21908 63625 15982
	0.86506 33666 88985	0.14945 13491 50581
	0.97390 65285 17188	0.06667 13443 08648

In this exercise we use Gaussian integration to find the value of:

$$I = \int_{-1}^{1} x^2 \cos(x)dx$$

To make the worksheet more versatile, we will code a user-defined function for x^2 cos(x); in this way we will be able to perform Gaussian integrations on other functions merely by editing the user-defined function. We will increase the number of terms until we have an approximate value to four decimal places.

(a) Open CHAP11.XLS and invoke the VBE. Insert another module on the CHAP10 project on which to code the following function:

```
Function gaussfunc(x)
        gaussfunc = x ^ 2 * Cos(x)
    End Function
```

(b) Return to the workbook and on Sheet6 enter the text shown in A1:E3 of Figure 11.7.

(c) Select the column heading B:D and with Format|Column|Width set the width to 40.

(d) Enter the values shown in A4:A16 and in B4:B20. It is safer to use =5/9 in B7 and =8/9 in B8 rather than numerical values.

	A	B	C	D	E
1	Gaussian Integration				
2					
3	Terms	Weight	Point	Term	Approx
4	2	1	0.577350269	0.279303943	
5		1	-0.577350269	0.279303943	0.5586078851
6					
7	3	0.555555556	0.774596669	0.238234398	
8		0.888888889	0	0	
9		0.555555556	-0.774596669	0.238234398	0.4764687953
10					
11	4	0.652145155	0.339981044	0.071064922	
12		0.347854854	0.861136312	0.168076458	
13		0.347854854	-0.861136312	0.168076458	
14		0.652145155	-0.339981044	0.071064922	0.4782827589
15					
16	5	0.478628671	0.538469231	0.119140142	
17		0.236926885	0.906179846	0.119993421	
18		0.568888889	0	0	
19		0.236926885	-0.906179846	0.119993421	
20		0.478628671	-0.538469231	0.119140142	0.4782671248

Figure 11.7

(e) Enter the values shown in C4:C16. For C4 you may wish to use =1/SQRT(3) and the negative of this in C5. Similarly for C7 use =SQRT(3/5) and the negative of this in C9.

(f) In D4 enter =B4*Gaussfunc(C4). Copy this down to D20. Delete D6, D10 and D15.

(g) Move to E6 and click on the Autosum button. Drag over D4:D5 to give the formula =SUM(D4:D5). Repeat this operation for the three other approximations.

We can see that the four-point and the five-point approximations agree to four decimal places, so our task is complete. Save the workbook.

Exercise 7: Monte Carlo Techniques

There is no mathematical advantage to performing a numerical integration using a Monte Carlo technique. However, it does provide a simple way to illustrate a Monte Carlo calculation. A large worksheet would be needed to model a true stochastic process. The author's web page has a workbook demonstrating Brownian motion.

Consider a circle inscribed within a square with sides of l units. The radius (r) of the circle will be $l/2$. A large number (N) of darts are randomly thrown at the diagram and the number (C) that fall within the circle are counted. If the throwing was truly random then:

$$\frac{\text{Number of darts in circle}}{\text{Total number of darts}} = \frac{\text{Area of circle}}{\text{Area of square}}$$

or

$$\frac{C}{N} = \frac{\pi r^2}{l^2} = \frac{\pi}{4}$$

Hence the value of π may be approximated from a simple dart-throwing experiment. We will use this Monte Carlo method to get an approximate value of the integral

$$I = \int_0^{10} (-x^3 + 10x^2 + 5x)dx$$

As in previous exercises, we have chosen an integral which can be solved analytically so as to be able to evaluate our result.

(a) On Sheet7 of CHAP11.XLS begin by constructing the table shown in H1:I12 of Figure 11.8. The formula in I2 is =-H2^3+10*H2^2+5*H2. This is copied down to row 12. Make a chart of the data.

Our curve is enclosed by a 10 by 200 rectangle. We will use the RAND function to generate two random values from which we will find the position of the dart. The function returns a value from 0 to 1, so for the x value we will use RAND()*10 and for the y value RAND()*200.

(b) In A3 enter =RAND()*10 and in B3 enter =RAND()*200. Copy these down to row 1002. Your values will not be the same as in the figure. You may see a message such as *Calculation Cells 49%* on the status bar. If your PC reacts slowly, use Tools|Options and on the *Calculations* tab set the method of calculation to manual. A status bar message of *Calculate* will remind you when to press F9 to recalculate the worksheet.

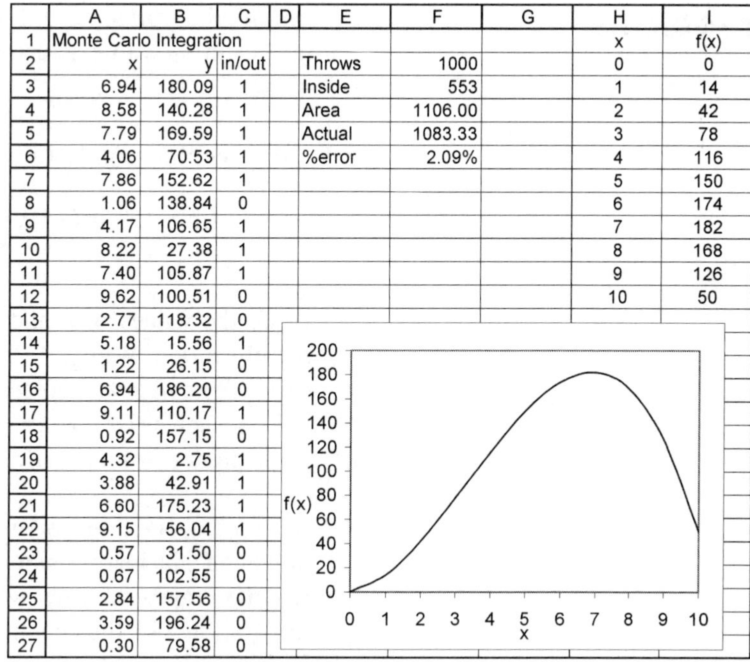

	A	B	C	D	E	F	G	H	I
1	Monte Carlo Integration							x	f(x)
2	x	y	in/out		Throws	1000		0	0
3	6.94	180.09	1		Inside	553		1	14
4	8.58	140.28	1		Area	1106.00		2	42
5	7.79	169.59	1		Actual	1083.33		3	78
6	4.06	70.53	1		%error	2.09%		4	116
7	7.86	152.62	1					5	150
8	1.06	138.84	0					6	174
9	4.17	106.65	1					7	182
10	8.22	27.38	1					8	168
11	7.40	105.87	1					9	126
12	9.62	100.51	0					10	50
13	2.77	118.32	0						
14	5.18	15.56	1						
15	1.22	26.15	0						
16	6.94	186.20	0						
17	9.11	110.17	1						
18	0.92	157.15	0						
19	4.32	2.75	1						
20	3.88	42.91	1						
21	6.60	175.23	1						
22	9.15	56.04	1						
23	0.57	31.50	0						
24	0.67	102.55	0						
25	2.84	157.56	0						
26	3.59	196.24	0						
27	0.30	79.58	0						

Figure 11.8

(c) The formula in C3 is =IF(B3>-A3^3+10*A3^2+5*A3,0,1). This returns 0 when the dart has fallen above the curve, and 1 otherwise. Copy it down to C1002.

(d) The formulas in column F are:

F2: =COUNT(A3:A1002)	Total darts thrown
F3: =SUM(C3:C1002)	Darts inside the curve
F4: =2000*F3/F2	Area under the curve
F5: =-(1/4)*10^4+(10/3)*10^3+(5/2)*10^2	Analytical result
F6: =(F4-F5)/F5	Relative error

(e) Repeatedly press F9 to recalculate the worksheet even if you have not set the calculation method to manual. A new set of random numbers is generated each time leading to a new value in F5. The error seems to lie within a range of ±5% of the analytical value. Adding more random numbers would help. Alternatively, we could record the result for, say, 20 recalculations and take an average. This could be done automatically using a VBA subroutine.

(f) If you have set the calculation method to manual, reset it to automatic. Save the worksheet.

Problems

1. Modify the worksheet of Exercise 1 to use 20 strips. Note how the resulting value is a better approximation to the known value.

2.* Using the Simpson ⅓ rule with 20 strips, show that:

$$\int_0^1 x^p (\ln x)^q \, dx = (-1)q \frac{q!}{(p+1)^{q+1}}$$

Place the values of p and q in named cells so that they are readily varied. With $p = 2$, $q = 1$, the value from the Simpson rule will be in good agreement with the exact value. Try other values, such as ($p = 8$, $q = 4$), ($p = 1$, $q = 5$), etc. Explain why some pairs of values yield poor approximations.

3.* Set up a worksheet to find the approximate integral of the tabulated data below using (a) the trapezoid rule with all strips, (b) the trapezoid rule with alternate strips, (c) Romberg integration with the data from (a) and (b), and (d) the Simpson ⅓ rule. Which of your answers do you believe to be more exact?

x	y	x	y	x	y
1.00	1.609438	1.35	1.761730	1.70	1.930071
1.05	1.629731	1.40	1.785070	1.75	1.954799
1.10	1.650580	1.45	1.808699	1.80	1.979621
1.15	1.671943	1.50	1.832581	1.85	2.004516
1.20	1.693779	1.55	1.856689	1.90	2.029463
1.25	1.716048	1.60	1.880991	1.95	2.054444
1.30	1.738710	1.65	1.905460	2.00	2.079442

4.* Find the approximate value of the area under the curve represented by the data in the following table. Suggest a method to check your result and perhaps find a better approximation.

x	3.00	3.60	4.00	4.70	5.50	6.25	7.00	7.50
$f(x)$	1.17	1.02	0.90	0.63	0.59	0.63	0.66	0.79
x	8.60	9.00	9.40	10.00	10.45	10.85	11.25	
$f(x)$	1.20	1.45	1.65	2.05	2.26	2.62	3.00	

5. The accompanying table contains an even number of data pairs. We cannot use Simpson's ⅓ rule to find the approximate area under the curve since there are an odd number of strips. Use Simpson's ⅜ rule for the first three strips and the ⅓ rule for the remainder.

x	y	x	y
0.000	0.25000	0.500	0.26667
0.100	0.25063	0.600	0.27473
0.200	0.25253	0.700	0.28490
0.300	0.25575	0.800	0.29762
0.400	0.26042	0.900	0.31348

6.* Using Simpson's ⅓ rule, find the area of the region R bounded by $y = 4x^2$ and $y = x^4$ for positive x values. How well does your result compare to the exact value?

12
Differential Equations

Concepts

Differential equations occur in many physical problems. Let us look at some simple examples.

(i) A body falling through the air is subjected to two forces: gravity acting downwards and air resistance acting upwards. The first force is constant but the second is proportional to the body's velocity. This gives rise to the first-order differential equation

$$m\frac{dv}{dt} = g - kv^2 \tag{12.1}$$

(ii) Consider the chemical reaction A + B → C where the rate of reaction is proportional to the concentration of A and to the concentration of B. Let x be the amount of A and B reacted at time t, and let the initial concentration of A and B be a and b, respectively These quantities will be related by

$$\frac{dx}{dt} = k_2(a-x)(b-x) \tag{12.2}$$

(iii) The equation of motion for a harmonic oscillator is

$$\frac{d^2x}{dt^2} + \omega x = 0 \tag{12.3}$$

Equations 12.1 and 12.2 are examples of first-order differential equations while Equation 12.3 is of second order. The equations in these examples may readily be solved by analytical means. The next equation is also first order but its solution is rather more difficult to arrive at:

$$\frac{dy}{dx} = \frac{x+y}{x-y} \tag{12.4}$$

When a differential equation is difficult or impossible to solve analytically we may use numerical methods to find approximate solutions.

Consider the simple equation $dv/dt = g$ for a falling body when air resistance is ignored. This integrates to give $v = gt + c$ where c is the integration constant and g is a constant of known value. Thus we do not have a unique solution, since any value of c will satisfy the differential equation. By inspection of the solution we see that c is the value of v when t equals zero. We need to know this value in order to uniquely solve the equation. In general, to solve $dy/dx = f(x,y)$ over the x range $[a, b]$, we need to know the value of $y(a)$ which is called the *initial value.* Problems of this type are called *initial value* problems. With second-order differential equations two integration constants arise. For an initial value problem we need to know the initial value of the two values of the dependent variables. Alternatively, the problem may be defined by specifying some conditions at one value of x and others at another value of x. Such problems are called *boundary value* problems.

Exercise 1: Euler's Method

Euler developed a method for finding the approximate solution to initial value problems. Let the differential equation to be solved have the form of Equation 12.5 and let the initial value of y be y_0.

$$\frac{dy}{dx} = y' = f(x,y) \tag{12.5}$$

Let the solution (i.e. the integral of Equation 12.5) have the form of Equation 12.6.

$$y = g(x,y) \tag{12.6}$$

Consider the two curves in Figure 12.1. From Equation 12.5, we may calculate any value $y'_1, y'_2, \ldots, y'_n$. We already know the value of y_0 – the initial value. Our task is to find values for $y_1, y_2, \ldots, y_n$. Integration of Equation 12.5 from x_0 to x_1 yields Equation 12.7.

$$y_1 = y_0 + \int_{x_0}^{x_1} f(x_0, y_0)dx \tag{12.7}$$

The second term on the right is the area under the curve $f(x,y)$ between the two x values. Euler approximated this to the area of the rectangle defined by y'_0, y'_1, x_0 and x_1. The approximate value of y_1 is then given by Equation 12.8, or by Equation 12.9 when the x increment is represented by h.

$$y_1 = y_0 + (x_1 - x_0)f(x_0, y_0) \tag{12.8}$$

$$y_1 = y_0 + hf(x_0, y_0) \tag{12.9}$$

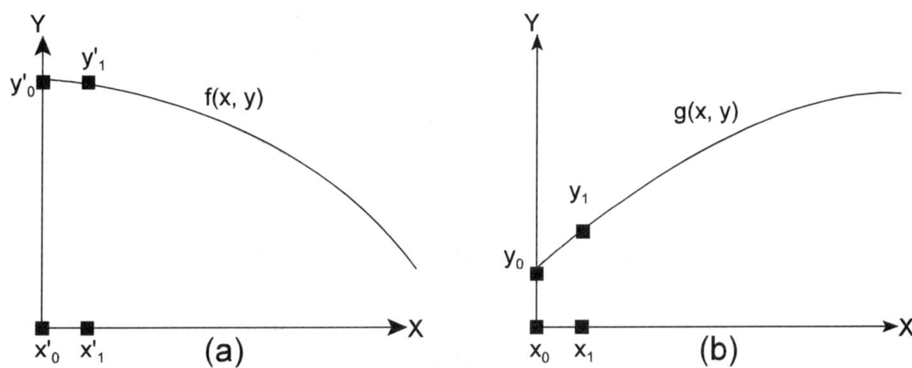

Figure 12.1

Having found an approximation for y_1, we may now find an approximation for y_2

$$y_2 = y_1 + hf(x_1, y_1) \tag{12.10}$$

In general, the value of the approximation at one point is found from the previous one using

$$y_{n+1} = y_n + hf(x_n, y_n) \tag{12.11}$$

In this exercise, Euler's method is used to find an approximate solution of the differential equation $dy/dx = xy$, with the initial value $y(0) = 1$. The approximation is compared to the analytical solution $y = \exp(x^2/2)$.

	A	B	C	D	E	F
1	Euler's Method					
2	Differential equation y'=xy with boundary condition y(0)=1					
3						
4	i	x	y	h*f(x,y)	exact	
5	0	0	1.00000	0.00000	1.00000	
6	1	0.1	1.00000	0.01000	1.00501	
7	2	0.2	1.01000	0.02020	1.02020	
8	3	0.3	1.03020	0.03091	1.04603	
9	4	0.4	1.06111	0.04244	1.08329	
10	5	0.5	1.10355		1.13315	

Figure 12.2

(a) Open a new workbook. Enter the text shown on A1:E4 of Figure 12.2.

(b) Enter the series of values in A5:A10 and B5:B10.

(c) In C5 enter the value 1.0. This is the initial condition $y(0) = 1$ and corresponds to the first term on the right of Equation 12.11 when $n = 0$.

(d) In D5 enter =0.1*(B5*C5). This corresponds to the second term on the right in Equation 12.11 when $n = 0$. The parentheses are not essential here but are used to make it clear that we are computing the value *h*function.*

(e) In C6 enter =C5 + D5. This corresponds to the left-hand side of Equation 12.11 and calculates the first approximate value for *y*.

(f) Copy the formula from D5 to D6.

(g) Copy the cells C6:D6 down to row 10. This computes the successive *y* approximations. Since we shall not be using the last value of *h*f(x,y)*, delete D10.

(h) So that we can compare our approximations in the C column with the exact solution, in E5 enter =EXP(B5^2 / 2) and copy this down to E10.

(i) Save the workbook as CHAP12.XLS.

Clearly, our answer in the C column is not in very good agreement with the exact values in the E column. A better approximation may be obtained by reducing the size of *h*, the increment for the *x* values. Figure 12.3 graphs the exact solution, the approximations with $h = 0.1$ and the approximations with $h = 0.05$. This shows (i) that the deviation from the exact values increases with each iteration of Equation 12.11 as expected, and (ii) decreasing the size of *h* significantly improves the solution values.

(j) Modify your worksheet to find the approximations of this differential equation for *x* values from 0 to 0.5 with steps of 0.025. Save the workbook.

In this example we have used the 'crude' Euler method in which the integral in Equation 12.7 was approximated to the area of a rectangle. In the modified Euler method it is approximated to a trapezoid. Compared to the original Euler method, this requires fewer calculations for comparable accuracy. We shall not examine the modified method. The next exercise uses a more modern method.

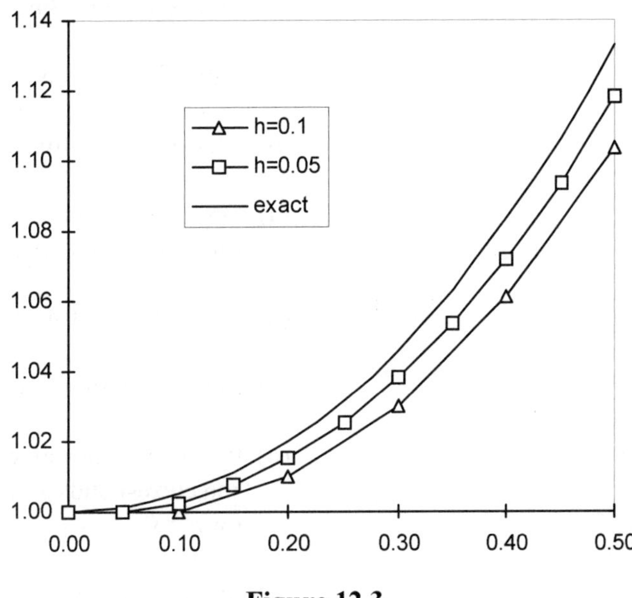

Figure 12.3

Exercise 2: The Runge–Kutta Method

The Runge–Kutta methods are similar in concept to the Euler method in that each approximation for y is based on the previous value. These mathematicians developed a number of algorithms to solve differential equations. We shall use the fourth-order Runge–Kutta method, the derivation of which is beyond the scope of this book. The iterative formula is given in Equation 12.12. This is sometimes called the *Kutta–Simpson* formula since, when the right-hand side of the differential equation is a function of x alone, it reduces to Simpson's one-third rule.

$$\begin{aligned} y_{n-1} &= y_n + \tfrac{1}{6}(k_1 + 2k_2 + 2k_3 + k_4) \\ k_1 &= hf(x_n, y_n) \\ k_2 &= hf(x_n + \tfrac{1}{2}h, y_n + \tfrac{1}{2}k_1) \\ k_3 &= hf(x_n + \tfrac{1}{2}h, y_n + \tfrac{1}{2}k_2) \\ k_4 &= hf(x_n + h, y_n + k_3) \end{aligned} \tag{12.12}$$

In Exercise 4 of Chapter 13 a user-defined function is developed to perform the Runge–Kutta computation.

This may look somewhat formidable so let us see how we can put it into a worksheet. We have seen in Exercise 1 how to evaluate the equivalent to k_1; this is the value of the differential function for various x and y values. The second parameter, k_2, is similar except that the x value is incremented by h while the y value is incremented by k_1. Each parameter increments y by a multiple of the parameter preceding it.

In the previous exercise we solved $dy/dx = xy$ with the initial value $y(0) = 1$ using Euler's method. Here we solve the same problem using the Runge–Kutta method so that we may compare the results.

(a) Open the workbook CHAP12.XLS and make Sheet2 active. Enter the text shown on A1:H5 of Figure 12.4.

(b) Name the cell C4 as *h*.

(c) Enter the series of values in A6:A11 and B6:B11.

(d) In C6 enter the value 1.0. This is the initial condition $y(0) = 1$ and corresponds to the first term to the right in Equation 12.12 for $n = 0$.

(e) Enter these formulas to compute the k parameters:
D6: =h*(B6 * C6)
E6: =h*((B6 + h/2) * (C6 + D6/2))
F6: =h*((B6 + h/2) * (C6 + E6/2))
G6: =h*((B6 + h) * (C6 + F6))

	A	B	C	D	E	F	G	H
1	Runge-Kutta Method							
2	Differential equation y' = x*y with boundary condition y(0)=1							
3								
4		h =	0.1					
5	i	x	y	k1	k2	k3	k4	Error
6	0	0.0	1.00000	0.00000	0.00500	0.00501	0.01005	0.000E+00
7	1	0.1	1.00501	0.01005	0.01515	0.01519	0.02040	-2.607E-11
8	2	0.2	1.02020	0.02040	0.02576	0.02583	0.03138	-2.684E-10
9	3	0.3	1.04603	0.03138	0.03716	0.03726	0.04333	-1.023E-09
10	4	0.4	1.08329	0.04333	0.04972	0.04987	0.05666	-2.854E-09
11	5	0.5	1.13315					-6.949E-09

Figure 12.4

(f) In C7 enter =C6 + (1/6)*(D6 + 2*E6 + 2*F6 + G6) to compute the first approximation. Compare this formula with Equation 12.12.

(g) Copy the cells D6:G6 to D7:G7.

(h) Copy C7:G7 down to row 10 and copy C10 to C11. This computes the successive *y* approximations.

(i) So that we compare our approximations in the C column with the exact solution, in H5 enter =C6 - EXP(B6^2 / 2). Format this to display three decimal places in scientific notation and copy this down to H11.

(j) Save the workbook CHAP12.XLS.

For the equation *dy/dx* = *xy*, the Runge–Kutta method is clearly far superior to Euler's method. It may be shown that this is true for all equations.

Exercise 3: Solving with a User-defined Function

Any of the Copy and Paste methods may be used to duplicate the worksheet. Alternatively, hold down Ctrl and drag the tag of the sheet to be copied to the right. When Ctrl is released, a copy of the sheet is inserted into the workbook.

In the previous exercise we solved *dy/dx* = *xy*. We would need to make many edits to the worksheet to solve for another equation *dy/dx* = *f*(*x*,*y*). If we put the function *f*(*x*,*y*) in a module, we need edit only the module (and the initial value) to change our worksheet.

In this exercise we find the values of *y* which satisfy the equation *dy/dx* = 1/(*x* + *y*) with the initial value *y*(0) = 2. We use *x* values from 0 to 1.0 in increments of 0.2. The completed worksheet (Figure 12.5) is very similar to that in the previous exercise. To save time, you may wish to copy that worksheet to a new sheet and edit it to agree with the instructions below.

	A	B	C	D	E	F	G
1	Runge-Kutta Method						
2	Solving dy/dx = f(x,y) with a user-defined function						
3							
4	increment	h	0.2				
5	initial x value	x0	0				
6	initial y value	y0	2				
7	i	x	y	k1	k2	k3	k4
8	0	0.0	2.00000	0.10000	0.09302	0.09317	0.08722
9	1	0.2	2.09327	0.08721	0.08207	0.08216	0.07766
10	2	0.4	2.17549	0.07766	0.07368	0.07374	0.07019
11	3	0.6	2.24927	0.07019	0.06702	0.06705	0.06418
12	4	0.8	2.31636	0.06418	0.06157	0.06159	0.05921
13	5	1	2.37797				

Figure 12.5

(a) Open the VBE and insert a module on the CHAP12 project. Enter this function on the module sheet:

```
Function RKfunc(x, y)
    RKfunc = 1 / (x + y)
End Function
```

(b) On Sheet3 enter the text and values shown in A1:G7 of Figure 12.5. Select the range B4:C6 and use Insert|Name|Create to name the cells C4:C6.

(c) Enter the series of values in A8:A13.

(d) Enter these formulas:

B8:	=X0	The initial x value
C8:	=Y0	The initial y value
D8:	=h*rkfunc(B8, C8)	The k parameters
E8:	=h*rkfunc ((B8 + h/2), (C8 + D8/2))	
F8:	=h*rkfunc ((B8 + h/2), (C8 + E8/2))	
G8:	=h*rkfunc ((B8 + h), (C8 + F8))	

(e) In B9 enter =B8+h to increment x.

(f) In C9 enter =C8 + (1/6)*(D8 + 2*E8 + 2*F8 + G8) to compute the first approximation for y.

(g) Copy D8:G8 to line 9. In row 9, we have the second y value, and the k values needed to compute the third y value.

(h) Copy B9:G9 down to line 13 to compute the successive y values. Save the workbook.

If your values do not agree with Figure 12.5 you need to check the function in the module and the formulas on the worksheet. Remember that formulas can be displayed with Ctrl+`. To check the function, move to a blank cell such as A20 and enter =rkfunc(3,1). This should return the value 0.25.

(i) Now that you have solved one equation, modify the module sheet and the values in named cells of the worksheet to solve the equation $dy/dx = x^2 + y$ with the initial value $y(1) = 1$. Find the value of y when $x = 1.5$ using first $h = 0.1$, then $h = 0.01$. You will need to extend the worksheet in the second case. The analytical result to eight decimal places is $y(1.5) = 2.64232762$.

Simultaneous and Second-order Differential Equations

Consider a pair of simultaneous equations having the form:

$$\begin{aligned} y' &= g(x, y, z) \\ u' &= f(x, y, z) \end{aligned} \tag{12.13}$$

The Runge–Kutta formulas for these equations are given in Equation (12.14).

$$\begin{aligned}
y_{n-1} &= y_n + \tfrac{1}{6}(k_1 + 2k_2 + 2k_3 + k_4) \\
u_{n-1} &= u_n + \tfrac{1}{6}(q_1 + 2q_2 + 2q_3 + q_4) \\
k_1 &= hg(x_n, y_n, u_n) \\
q_1 &= hf(x_n, y_n, u_n) \\
k_2 &= hg(x_n + \tfrac{1}{2}h, y_n + \tfrac{1}{2}k_1, u_n + \tfrac{1}{2}q_1) \\
q_2 &= hf(x_n + \tfrac{1}{2}h, y_n + \tfrac{1}{2}k_1, u_n + \tfrac{1}{2}q_1) \\
k_3 &= hg(x_n + \tfrac{1}{2}h, y_n + \tfrac{1}{2}k_2, u_n + \tfrac{1}{2}q_2) \\
q_3 &= hf(x_n + \tfrac{1}{2}h, y_n + \tfrac{1}{2}k_2, u_n + \tfrac{1}{2}q_2) \\
k_4 &= hg(x_n + h, y_n + k_3, u_n + q_3) \\
q_4 &= hf(x_n + h, y_n + k_3, u_n + q_3)
\end{aligned} \tag{12.14}$$

Equations of higher order than first may be solved by transforming them into sets of simultaneous equations. For example, to solve $y'' = ay' + by + c$, we make the substitution $y' = u$. The introduction of the auxiliary variable u allows us to write the second-order equations as two simultaneous equations:

$$\begin{aligned} y' &= u \\ u' &= au + by + c \end{aligned} \tag{12.15}$$

Comparing Equations 12.14 and 12.15, we see that g is a function only of u and is a very simple function: it has the value of u. This simplifies the k terms in Equation 12.14:

$$\begin{aligned}
k_1 &= h(u_n) \\
k_2 &= h(u_n + \tfrac{1}{2}q_1) \\
k_3 &= h(u_n + \tfrac{1}{2}q_2) \\
k_4 &= h(u_n + q_3)
\end{aligned} \tag{12.16}$$

Exercise 4: Solving a Second-order Equation

In this exercise we apply the equations developed above to solve:

$y'' = y' + y = \sin(x)$

with boundary conditions $y(0) = 0$ and $y'(0) = 0$

Our task is to obtain approximate values of y and y' when $x = 1$.

With the substitution $y' = u$, we get a pair of equations:

$y' = u$ initial value $y(0) = 0$

$u' = \sin(x) - y - u$ initial value $u(0) = 0$

Comparing these with Equation 12.13, we see that $g = u$, so we will use the simplified k values of Equation 12.16. We also see that $f = \sin(x) - y - u$.

The function f is referenced in each of the q terms, so it will be more convenient to use a module function. Furthermore, by changing the module you will be able to use the same worksheet for another function.

(a) With CHAP12.XLS open, go to the VBE and insert a new module. For this exercise, code the function:

```
Function f(x, y, u)
  f = Sin(x) - y - u
End Function
```

(b) Move to Sheet4 and enter the text and values shown in A1:K6 of Figure 12.6.

(c) Select A3:D4 and name the cells in row 4.

	A	B	C	D	E	F	G	H	I	J	K
1	Second-order differential equation						y" + y' + y = sin(x)				
2							y"(0) = y(0) = 0				
3	xinit	yinit	uinit	h							
4	0	0	0	0.2							
5											
6	x	y	u	k1	q1	k2	q2	k3	q3	k4	q4
7	0.0	0.000	0.000	0.000	0.000	0.000	0.020	0.002	0.018	0.004	0.036
8	0.2	0.001	0.019	0.004	0.036	0.007	0.051	0.009	0.049	0.014	0.062

Figure 12.6

(d) The formulas in row 7 are shown below.

A7: =xinit

B7: =yinit

C7: =uinit

D7: `=h*C7`
E7: `=h*f(A7,B7,C7)`
F7: `=h*(C7+E7/2)`
G7: `=h*f(A7+h/2,B7+D7/2,C7+E7/2)`
H7: `=h*(C7+G7/2)`
I7: `=h*f(A7+h/2,B7+F7/2,C7+G7/2)`
J7: `=h*(C7+I7)`
K7: `=h*f(A7+h,B7+H7,C7+I7)`

(e) The formulas in row A8:C8 are shown below. Those in columns D to K may be copied from the row above.
A8: `=A7+h`
B8: `=B7+(D7+2*F7+2*H7+J7)/6`
C8: `=C7+(E7+2*G7+2*I7+K7)/6`
D8: `=h*C8`

(f) Copy row 8 down to row 12 to get a final value of $x = 1.0$. The results should be $y(1.0) = 0.119394$ and $y'(1.0) = 0.307960$.

(g) Try other values of h such as 0.1 and 0.05 to see if the approximations converge. You will need to expand the table to have $x = 1.0$ in the final row.

Exercise 5: The Simple Pendulum

The equation of motion for a simple pendulum of length L is

$$\frac{d^2\theta}{dt^2} - \frac{g}{L}\sin(\theta) = 0$$

Most textbooks consider a pendulum which starts with a small displacement and use the approximation $\sin(\theta) \approx \theta$. Our approximation will be to use the Runge–Kutta method to solve this second-order differential equation to show how the angle and angular velocity change with time. We will model a 0.75 metre pendulum which is started with a displacement of 0.8 radians from the perpendicular.

As before, we start with the substitution $d\theta/dt = u$, giving:

$\theta' = u$ $\qquad$ $\theta\,(t = 0) = 0.8$

$u' = -(g\,/\,L)\sin(\theta)$ $\qquad$ $u\,(t = 0) = 0$

(a) On the same module used for Exercise 4, code the Pend function

```
Function Pend(L, angle )
        g = 9.8
        Pend = (-g/L)  *  Sin(angle)
End Function
```

The parentheses around *g/L* are not essential but help in reading the formula.

(b) On Sheet5 enter the text and values shown in A1:K6 of Figure 12.7.

	A	B	C	D	E	F	G	H	I	J	K
1	Pendulum										
2											
3	InitTime	InitAngle	InitVel	Length	h						
4	0	0.8	0	0.75	0.1						
5											
6	Time	Angle	Velocity	k1	q1	k2	q2	k3	q3	k4	q4
7	0.0	0.800	0.000	0.000	-0.937	-0.047	-0.937	-0.047	-0.916	-0.092	-0.894
8	0.1	0.753	-0.923	-0.092	-0.894	-0.137	-0.849	-0.135	-0.827	-0.175	-0.758

Figure 12.7

(c) The formulas needed in row 7 start with:
A7: =InitTime
B7: =InitAngle
C7: =InitVel
D7: =h*C7
E8: =h*Pend(Length, B7)
F7: =h*(C7+E7/2)
G8: =h*Pend(Length, B7+D7/2)

(d) From what you learnt in Exercise 4, complete the formulas in row 7 and in row 8. Copy row 8 down to row 37.

(e) Make a chart showing how the angle and the velocity vary with time.

Problems

1.* Solve the first-order differential equation $y' = -xy^2$ with $y(0) = 1$ with steps of $h = 0.2$ to estimate the value of $y(1)$ using (a) Euler's method and (b) the Runge–Kutta method.

2. In this chapter we have examined two methods of solving differential equations. There are many more. One of these is the *midpoint method* which uses:

$$y_{n+1} = y_n + hf(x_n + \tfrac{1}{2}h, y_n + \tfrac{1}{2}hf(x_n, y_n))$$

Develop a worksheet to use this method to solve $y' = xy$ at $x = 0.5$ when the initial condition is $y(0) = 1$. We solved the same problem in Exercises 1 and 2. Which of the three methods gives the more accurate result with the same h increment?

3. An analysis of a certain electric circuit shows it obeys the equations:

$$dI_1/dt = 12I_1 + 10I_2 + 10$$
$$dI_2/dt = 10I_1 - 20I_2$$

where $I_1(0) = I_2(0) = 0$.

Use the Runge–Kutta method to approximate $I_1(t)$ and $I_2(t)$ at $t = 0$ to 0.5 seconds in 0.1 second intervals.

4. A rocket has a mass of 2000 kg of which 1500 kg is fuel. It burns the fuel at the rate of 25 kg/s and develops a thrust of 5000 N. Construct a worksheet to find how the velocity varies with time. What will be the height when the fuel is all consumed?

13 Modelling II

Concepts

In this chapter we use what we have learnt in the last three chapters to model some practical problems.

Exercise 1: The Four-bar Crank: Using Solver

In this exercise we examine an engineering mechanism used to generate a complex motion from a simple rotational motion.

Description
The four-bar mechanism (see Figure 13.1) consists of three movable links (*a*, *b* and *c*) and a fixed link *d*. The link *a* is rotated causing link *c* to rotate.

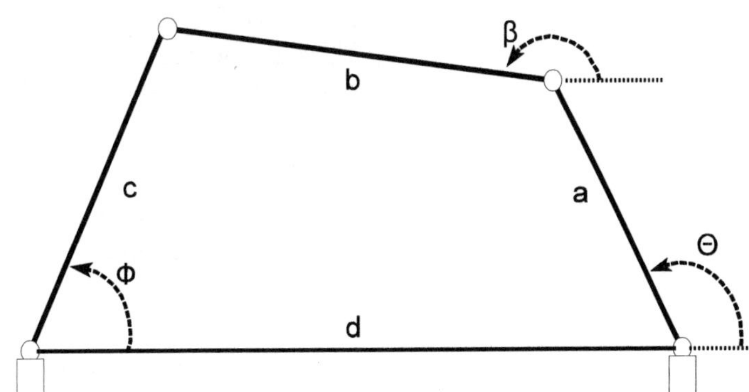

Figure 13.1

Objective
To find how the output angle (ϕ) depends upon the input angle (θ).

Assumptions
For the quadrilateral formed by the four links, the algebraic sum of the vertical component and the algebraic sum of the horizontal component must equate to zero. This gives the two equations:

$$a\sin\theta + b\sin\beta - c\sin\phi = 0$$
$$a\cos\theta + b\cos\beta - c\cos\phi + d = 0$$

Adding the squares of these gives the Freudenstein equation:

$$R_1 - R_2\cos\phi + R_3 - \cos(\theta - \phi) = 0$$

where:

$R_1 = d/c$
$R_2 = d/a$
$R_3 = a^2 - b^2 + c^2 + d^2)/2ac$

Rather than attempt the difficult task of solving this to find ϕ in terms of θ, we will use Microsoft Excel's Solver to find the output angle for input angles in the range 0 to 360° in 5° steps.

	A	B	C	D	E	F	G	H
1	Four-bar Crank			Input Angle		Output Angle		Freudenstein's
2				Degrees	Radian	Degrees	Radian	Equation
3				0	0	57.3	1	6.29E-01
4	Crank lengths			5	0.087266	57.3	1	5.54E-01
5	a	1		10	0.174533	57.3	1	4.76E-01
6	b	2		15	0.261799	57.3	1	3.96E-01
7	c_	2		20	0.349066	57.3	1	3.14E-01
8	d	2		25	0.436332	57.3	1	2.30E-01
9	Ratio1	1		30	0.523599	57.3	1	1.47E-01
10	Ratio2	2		35	0.610865	57.3	1	6.33E-02
11	Ratio3	1.25		40	0.698132	57.3	1	-1.93E-02
12				45	0.785398	57.3	1	-1.01E-01
13	Solver Target			50	0.872665	57.3	1	-1.80E-01
14	SUMSQ	35.56002		55	0.959931	57.3	1	-2.56E-01
15				60	1.047198	57.3	1	-3.29E-01

Figure 13.2

(a) Enter the text shown in rows 1 and 2 of Figure 13.2 on Sheet1 of a new workbook.

The next stage is to enter the specifications for the four-bar crank and to compute the R values for the Freudenstein equation. It will be convenient to use named cells.

(b) Enter the text and values in A4:B8. Select A5:B8 and name the cells B5:B8.

(c) Enter the text in A9:A11. Enter the formulas:
B9: =d/c_
B10: =d/a
B11: =(a^2 -b^2 + c_^2 + d^2)/(2*a*c_)

(d) Select A9:B11 and name the cells in B9:B11.

We begin by solving the Freudenstein equation for two input angles, 0° and 5°.

(e) Enter 0 and 5 in D3 and D4, respectively.

(f) While we wish to have the input and output angles expressed in degrees, we shall also need the radian values to use the cosine function in the Freudenstein equation. So, enter the following formulas:

E3: =RADIANS(D3) Converts the input angle to radians
F3: =DEGREES (G3) Converts the output angle to degrees
G3: 1 The starting point for Solver
H3: =Ratio1*COS(E3) – Ratio2*COS(G3) + Ratio3 – COS(E3-G3)

Formatting F3 to show one decimal place improves the readability of the worksheet.

(g) Copy E3:H3 down to line 4.

(h) Invoke Solver (Tools|Solver). Click the *Reset All* button to restore the options to the default values: Max Time = 100 secs, Iterations = 100, Precision = 0.000001 and Convergence = 0.0001.

(i) Set the *Target Cell* to H3. Click the radio button *Equal to Value* and enter 0 in the value box. In the *By Changing Cells* enter G3. Click the *Solve* button. Solver should be given a value close to 0 in H3. The output angle in G3 should be approximately 0.72 radians and the corresponding value in degrees in F3 should be 41.4.

(j) Use Solver to find the output angle corresponding to an input angle of 5° in row 4. Remember to change the *Target* and *Changing* cells. The result should be 43.1°.

Do not be concerned that the values in column H are not identically zero but are in the range 1×10^{-5} to 1×10^{-7} depending on your computer system. We are modelling a real system; the components will have some slack so we do not need extreme precision.

We now expand the worksheet in preparation for finding the output angles corresponding to input angles in the range 0 to 360 in 5 degree intervals.

(k) Enter values of 1 in G3:G4. Select D3:H4 and copy the cells down to line 75. The value in D75 will be 360 – we have covered the full input range.

(l) Enter the text shown in A13:A14 and in B14 enter the formula =SUMSQ(H3:H75). This should result in the value 35.56.

(m) Use Solver with the following input parameters: *Target* B14, *Equal to Value of* 0 and *Changing Cell* G3:G75.

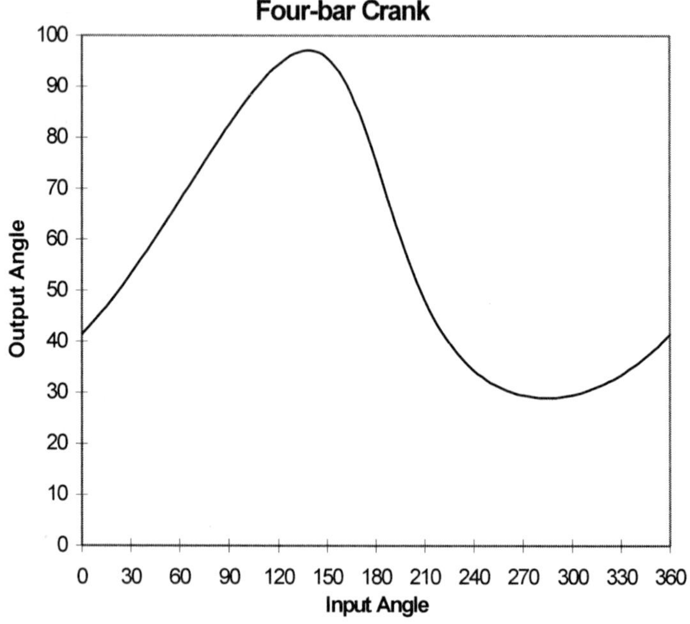

Figure 13.3

This is a very large problem so some additional notes on Solver are called for:

1. The progress of Solver may be monitored by observing the message in the status bar. You will see something like: *Trial Solution: 30 Set target cell: 1.87E–01.*

2. It is possible that Solver will report one of the following conditions: *The maximum time limit was reached: continue anyway?* or *The maximum iterations limit was reached; continue anyway?* If either message occurs, click *Continue*.

3. If at any time you wish to halt Solver, press Ctrl+Break.

In earlier editions of this book we solved this problem by using Solver four times – one for each quadrant of the input range. Modern Pentium computers have no trouble finding the solution to 73 equations.

When Solver has completed its task B14 will have a value of about 3×10^{-7}. There is too much data to absorb so it would be a good idea to make a chart. Figure 13.3 shows the results graphically.

You may wish to try other values for the lengths of the cranks. Bear in mind that the sum $a + b + c\_$ should be somewhat larger than d.

Exercise 2: Temperature Profile: Circular References

When the formula in a cell refers to its own address we say that there is a *circular reference* in the worksheet. Normally, we wish to avoid this condition but there are occasions when it is useful. One such use is explored in this exercise.

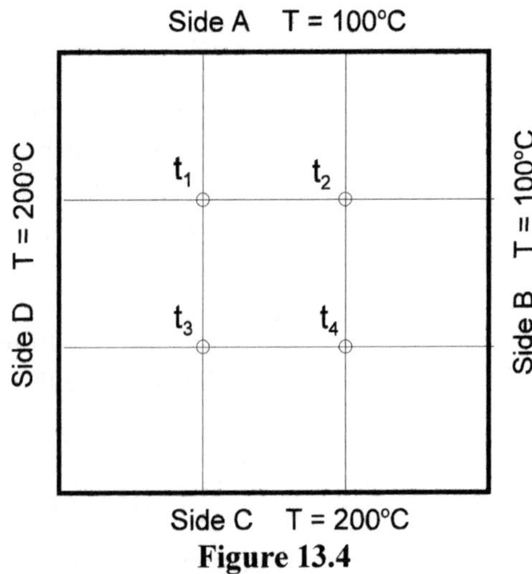

Figure 13.4

Description

The edges of a thin metal sheet are maintained at specified temperatures and the sheet is allowed to come to thermal equilibrium.

Objective

To compute the approximate temperatures at various positions on the plate.

Assumptions

The first assumption is that the two faces of the plate are thermally insulated. Thus there is no heat transfer perpendicular to the plate. The second assumption starts with the mean-value theory which states: if P is a point on a plate at thermal equilibrium and C is a circle centred on P and completely on the plate, then the temperature at P is the average value of the temperature on the circle. The calculations required to use this theory are formidable so we will use an approximation. We shall consider a finite number of equidistant points on the plate and use the discrete mean-value theory which states that the temperature at point P is the average of the temperatures of P's nearest neighbours.

The most convenient way to arrive at the equidistant point is to divide the plate using equally spaced vertical and horizontal lines. In Figure 13.4 two such lines have been drawn parallel to each axis. This gives four interior points for the calculation. With such a small number, the results will not be very accurate. However, the methodology is the same regardless of the number of points, and it is simpler to describe and test the method initially with four points.

The equations for the four interior points are:

$$\begin{aligned} t_1 &= (100 + t_2 + t_3 + 200)/4 \\ t_2 &= (100 + 100 + t_4 + t_1)/4 \\ t_3 &= (t_1 + t_4 + 200 + 200)/4 \\ t_4 &= (t_2 + 100 + 200 + t_3)/4 \end{aligned} \qquad (13.1)$$

We could use formulas equivalent to these in the worksheet but will opt to use the AVERAGE worksheet function.

	A	B
1	Temperature Profile	
2		
3		
4	SideA	100
5	SideB	100
6	SideC	200
7	SideD	200
8		
9	Flag	0
10		
11	T1	0
12	T2	0
13	T3	0
14	T4	0

Figure 13.5

	A	B
1	Temperature Profile	
2		
3		
4	SideA	100
5	SideB	100
6	SideC	200
7	SideD	200
8		
9	Flag	1
10		
11	T1	150
12	T2	125
13	T3	175
14	T4	150

Figure 13.6

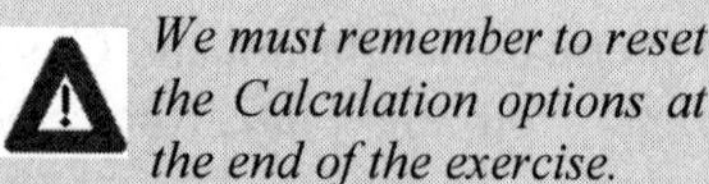
We must remember to reset the Calculation options at the end of the exercise.

(a) Move to Sheet2 of CHAP13.XLS. From the Tools menu select Options, and click on the *Calculation* tab. Set the calculation method to Manual. Click to the *Iteration* box to put a check mark in it and set the *Maximum Iterations* to 1.

(b) Enter the text shown in A1:A14 of Figure 13.5 and Select A4:B14 and name the cells in the B column. Although B11 is named T1, we shall need to refer to it as T1_ since T1 is a cell address.

(c) Enter the formulas shown below. Be careful to use commas not + signs between each item in the AVERAGE arguments.
B11: =IF(Flag=1, AVERAGE(SideA, T2_, T3_, SideD), 0)
B12: =IF(Flag=1, AVERAGE(SideA, SideB, T4_,T1_), 0)
B13: =IF(Flag=1, AVERAGE(T1_, T4_, SideC, SideD), 0)
B14: =IF(Flag=1, AVERAGE(T2_, SideB, SideC, T3_), 0)

The worksheet is now ready to use. Every time we enter a new value, the worksheet must be recalculated by pressing [F9] because we are using manual recalculation mode.

(d) Enter the value 1 in the Flag cell and press [F9]. The *T* values will change. Repeatedly tap [F9] and Microsoft Excel will repeat the calculation. After a number of iterations (20 or so), the values will cease to change and will be as shown in Figure 13.6. You may wish to check that the temperature of each internal point is the average of its nearest neighbours.

(e) If you wish to experiment with the *I* value do the following:

- (i) Enter 0 in the Flag cell and press [F9].
- (ii) Use Tools|Options to set the *Maximum Iterations* to 100.
- (iii) Enter a value of 1 in the Flag cell and press [F9].
- (iv) Repeatedly press [F9] until the *T* values remain constant. You will need to press [F9] far fewer times with *Maximum Iterations* set to 100.
- (v) Repeat steps (i) to (iv) to experiment with other values.

You may wish to make a note on the worksheet reminding yourself, or other users, to set *Manual Calculations* on next time the workbook is used.

Here is where we reset the Calculation options.

(f) Using Tools|Options, open the *Calculations* tab and reset Calculation to Automatic, uncheck the iterations box and set the *Maximum Calculations* to 100. Save the workbook.

You are encouraged to expand on the exercise to solve the same problem but using more internal points. Divide the plate into 36 squares and note how the temperatures of the points corresponding to those we have calculated come out with slightly different values.

Exercise 3: Temperature Profile: Matrix Method

In this exercise we model the same system as in Exercise 2 but use a matrix method to compute the temperatures. Again, we use only four internal points to facilitate discussion. A more accurate model is obtained using more points.

Equation 13.1 may be written in a more general form using variables *a*, *b*, *c* and *d* rather than numerical values. It may then be rearranged in the form:

$$\begin{aligned} t_1 &= (t_2 + t_3)/4 + (a + d)/4 \\ t_2 &= (t_4 + t_1)/4 + (a + b)/4 \\ t_3 &= (t_1 + t_4)/4 + (c + d)/4 \\ t_4 &= (t_2 + t_3)/4 + (b + c)/4 \end{aligned} \qquad (13.2)$$

To be able to use a matrix method each equation in Equation 13.2 must have the same form:

$$\begin{aligned} t_1 &= (0.00t_1 + 0.25t_2 + 0.25t_3 + 0.00t_4) + (a + d)/4 \\ t_2 &= (0.25t_1 + 0.00t_2 + 0.00t_3 + 0.25t_4) + (a + b)/4 \\ t_3 &= (0.25t_1 + 0.00t_2 + 0.00t_3 + 0.25t_4) + (c + d)/4 \\ t_4 &= (0.00t_1 + 0.25t_2 + 0.25t_3 + 0.00t_4) + (b + c)/4 \end{aligned} \qquad (13.3)$$

We may write this system of four equations as:

$$T = MT + B \qquad (13.4)$$

where:

$$T = \begin{bmatrix} t_1 \\ t_2 \\ t_3 \\ t_4 \end{bmatrix}, \quad M = \begin{bmatrix} 0 & 0.25 & 0.25 & 0 \\ 0.25 & 0 & 0 & 0.25 \\ 0.25 & 0 & 0 & 0.25 \\ 0 & 0.25 & 0.25 & 0 \end{bmatrix}, \quad B = \begin{bmatrix} (a+d)/4 \\ (a+b)/4 \\ (c+d)/4 \\ (b+c)/4 \end{bmatrix}$$

To solve Equation 13.4 we rewrite it as:

$$(I - M)T = B \qquad (13.5)$$

or

$$T = (I - M)^{-1}B \qquad (13.6)$$

From Equation 13.6 we note that, in the worksheet, we shall need a matrix *M* for the coefficients, an identity matrix *I* of the same rank as *M*, a matrix *B* for the temperatures of the sides, and a matrix *T* to hold the solutions.

	A	B	C	D	E	F	G	H	I
1	Temperature Profile			Using Matrix Method					
2									
3	SideA	SideB	SideC	SideD					
4	100	100	200	200					
5									
6									
7		M matrix				Inverse of I - M			
8	0	0.25	0.25	0		1.166667	0.333333	0.333333	0.166667
9	0.25	0	0	0.25		0.333333	1.166667	0.166667	0.333333
10	0.25	0	0	0.25		0.333333	0.166667	1.166667	0.333333
11	0	0.25	0.25	0		0.166667	0.333333	0.333333	1.166667
12									
13		I matrix				B matrix		T matix	
14	1	0	0	0		75		T1	150
15	0	1	0	0		50		T2	125
16	0	0	1	0		100		T3	175
17	0	0	0	1		75		T4	150
18									
19		I - M							
20	1	-0.25	-0.25	0					
21	-0.25	1	0	-0.25					
22	-0.25	0	1	-0.25					
23	0	-0.25	-0.25	1					

Figure 13.7

(a) On Sheet3 of CHAP13.XLS, enter the text and values shown in rows 1, 3 and 4 of Figure 13.7. Select A3:D4, name the cells A4:D4.

(b) Enter the text and values as shown in A7:D19.

If the status bar displays *Continue*, then you overlooked the last step in the previous exercise. You need to reset Excel for automatic calculations.

(c) In A20 enter the formula =A14 – A8. Copy this to A20:D23 to compute the elements of $[I - A]$.

(d) Enter the text in F7. Select F8:I11 and type the formula =MINVERS(A20:D23). Press Ctrl+⇧Shift+Enter to complete the array formula. This computes $[I - M]^{-1}$.

(e) Enter the text in F13. The formulas for this matrix are:
F14: =(SideA + SideD)/4
F15: =(SideA + SideB)/4
F16: =(SideC + SideD)/4
F17: =(SideB + SideC)/4

(f) Enter the text in H13:H17.

(g) All that remains is to multiply $[I - A]^{-1}$ by B. Select I14:I17, type the formula =MMULT(F8:I11, F14:F17). Press

Ctrl+Shift+Enter. The solutions in I14:I17 should agree with those found in the previous exercise.

(h) Save the workbook.

Exercise 4: Emptying the Tank

In this exercise we solve a simple differential equation using the Runge–Kutta method. In Chapter 12 we placed the terms needed for the Runge–Kutta approximation on the worksheet. In this exercise we use a user-defined function. In the subsequent exercise a function is used to iterate the approximation.

The analytical result is a parabolic function. Only values to the left of the minimum are meaningful.

The problem chosen has an analytical solution. You may wish to find it and compare the results from it with those found using the Runge–Kutta approximation.

Description
A cylindrical tank of diameter D has a short pipe of diameter d at the bottom. The tank is initially filled with water to a height h.

Objective
We wish to examine how changing the diameter of the pipe alters the rate of discharge of the tank.

Assumption
If the pipe is short, we may assume the rate of change of h is given by:

$$\frac{dh}{dt} = -\frac{d^2}{D^2}\sqrt{2gh}$$

Working in metric units, we shall use $g = 9.8 \text{ ms}^{-2}$.

We begin by developing a user-defined function to compute dh/dt for any value of h. Since we wish to vary the diameter of the pipe, we could treat d as a variable also. To make the function more general, it is better to make both d and D variables. The number of variables may be reduced if we write the equation as $dh/dt = -R^2\sqrt{2gh}$ where $R = d/D$. The quantity R is not really a variable; it is a parameter which we wish to vary from case to case. We pass it to the user-defined function in the same way as a true variable. The required function in given in Figure 13.8. This uses the VBA square root function SQR rather than the worksheet function SQRT.

```
'The function of the differential equation
Function tank(height, ratio)
   g = 9.8
   tank = -(ratio ^ 2) * Sqr(2 * g * height)
End Function
```

Figure 13.8

The function to perform the Runge–Kutta approximation is shown in Figure 13.9. Compare the *k* expressions with those in Equation 12.12. Since *t* does not appear to the right in the differential equation we are solving, there are no *x* terms in our *k* expressions. The *y* term of Equation 12.12 becomes the *height* term. What was called *h* in Equation 12.12, we call *incr* (short for *increment*) in our function. The *ratio* term has been added so that it may be passed to the *tank* function. Line 7 performs the final calculation. We must be careful in the worksheet to call this function with the *height*, *incr* and *ratio* arguments in the correct order.

The parentheses around 2*$k1$ and 2*$k2$ in line 7 are not required. However, because of the way VBA spaces, they were added to make the equation more readable .

```
1  'Function to compute Runge-Kutta approximation
2  Function RKapprox(height, incr, ratio)
3     k1 = incr * tank(height, ratio)
4     k2 = incr * tank(height + k1 / 2, ratio)
5     k3 = incr * tank(height + k2 / 2, ratio)
6     k4 = incr * tank(height + k3, ratio)
7     RKapprox = height + (k1 + (2 * k2) + (2 * k3) + k4) / 6
8  End Function
```

Figure 13.9

Having planned the user-defined functions needed, we are ready to start on the worksheet.

(a) Open CHAP13.XLS and invoke the VBE. Insert a module on which to code the two functions shown in Figures 13.8 and 13.9 on this module sheet.

(b) On Sheet 4, enter the values shown in A1:D7 of Figure 13.10. The formula in C7 is: =C6*0.01/C5. The 0.01 converts the pipe di0ameter to metres.

(c) Enter the text in B9:C9.

(d) Cells B10 and C10 contain the initial time and height values. In B10 enter the value 0 and in C10 enter the formula =C3.

	A	B	C	D
1	Tank Problem		Version 1	
2				
3	Initial height		1	
4	Time increment		0.1	
5	Tank diameter		1	metre
6	Pipe diameter		5	cm
7	Ratio of diameters		0.05	
8				
9		Time	Height	
10		0	1.000	
11		0.1	0.999	
12		0.2	0.998	
13		0.3	0.997	
14		0.4	0.996	
15		0.5	0.994	
16		0.6	0.993	
17		0.7	0.992	
18		0.8	0.991	
19		0.9	0.990	
20		1	0.989	

Figure 13.10

(e) On line 11 we compute the first approximation. Enter the formulas:
B11: =B10 + $C$4
C11: =RKapprox(C10, $C$4, $C$7)

The absolute references are used to enable us to copy the formulas.

(f) Hopefully, your worksheet returns the value 0.999 in C11. An error value of #NAME! means that the name of the function in the cell does not match that in the module. If C11 shows #VALUE!, check (i) that the arguments in the formula point to the correct value and (ii) that the RKapprox function is correctly coded.

(g) Save the workbook.

Clearly, the tank has discharged very little in 1 second. We cannot increase the value in C4 since this would cause the Runge–Kutta function to return inaccurate values. We must extend the worksheet if we wish to find the height at large time values.

(h) Copy B11:C11 down to row 110. Row 110 will give the height (0.892 m) at 10 seconds.

Our data extends over more than 100 rows and is too much to absorb. We need to make a summary.

(i) Enter the text shown in E1:F2 and the series shown in E3:E12 of Figure 13.11.

	E	F
1	Summary	
2	time	height
3	1	0.989
4	2	0.979
5	3	0.968
6	4	0.957
7	5	0.945
8	6	0.935
9	7	0.924
10	8	0.913
11	9	0.903
12	10	0.892

Figure 13.11

(j) In F3 enter the formula =VLOOKUP(E3, B10:C110, 2, TRUE). This function is used to do a 'lookup' on a vertical table such as the one in B10:C110. The first argument (E3) specifies that we wish to find in the first column (the function always searches the first column) of the table a value matching that in E3. The next argument gives the range of the table to be searched. The third argument (2) specifies that we wish to return the value in the second column of the table from the row in which a match was found. The final argument (TRUE) specifies that we want an exact, not an approximate, match. The function returns the value 0.989 since, on row 20 of the table, the value in the first column (B) has a value matching that in E3. The function then returns the value from the second column.

(k) Copy E3:F3 down to row 12. Save the workbook.

We may now vary the pipe diameter value and observe how the discharge changes. You may wish to make a graph of either B10:C110 or the summary data. Do not be misled into thinking that the height is linearly related to the time. This may be approximately true over the 10 second range of the data computed, but it will not hold true over a longer time range.

The worksheet is complete but is not really satisfactory. It has two major faults: (i) it gives discharge data for only the first 10 seconds and the tank is far from empty at this time, and (ii) while we can vary the pipe diameter, at any one time we can see data for only one pipe size. These matters are addressed in the next exercise.

Exercise 5: An Improved Tank Emptying Model

The worksheet we developed in the previous exercise had two faults. The second is readily addressed: use column D to hold the parameters and results for a second pipe diameter. The fact that the worksheet gives results for a limited time range could be addressed by extending the table further down the sheet but would result in a very large worksheet. In this exercise we investigate another way of achieving the same end.

Figure 13.12a shows our first attempt to design a function to call the RKapprox function *n* times.

We may call this function from the worksheet in, for example, C11 using the formula =NewHeight(C10, incr, ratio, 10) where C10 contains the starting value for the height of the water in the tank, *incr* is a name cell containing the increment for the Runge–Kutta calculation, and *ratio* is a named cell holding the value of the ratio of the diameters.

Line 4 invokes the Runge–Kutta function that we developed in the previous exercise. This will give *NewHeight* the value that we previously placed in C11. Rather than return that value to C11, the new function equates *OldHeight* to the new height value and calls the Runge–Kutta function again. In the previous exercise, this is equivalent to calling the Runge–Kutta function from C12 with C11's value for the height. This occurs 10 times and then the new function returns the height value to cell C11. Thus what took 10 rows in the previous worksheet is now done on one row.

The Runge–Kutta method is not accurate when *h* gets small compared to *incr* and may result in negative values which are meaningless in this model. We therefore modify the iteration function so that it stops before completing the *n* iterations if *h* is small. This is shown in Figure 13.12b.

```
'Function to make n calculations with RKapprox
Function NewHeight(OldHeight, incr, ratio, n)
    For j = 1 to n
        NewHeight = RKapprox(OldHeight, incr, ratio)
    OldHeight = NewHeight
    Next j
End Function
```

Figure 13.12a

We are now ready to develop the worksheet using the iteration function. We will also have columns to compute the water height for models having four different pipe and/or tank diameters.

(a) Open the VBE, insert a new module for CHAP13.XLS, and code the function shown in Figure 13.12b.

```
'Function to make n calculations with RKapprox
Function NewHeight(OldHeight, incr, ratio, n)
    For j = 1 to n
        If OldHeight < 0.0001 Then
        NewHeight = 0
            Exit For
        Else
            NewHeight = RKapprox(OldHeight, incr, ratio)
        OldHeight = NewHeight
        End If
    Next j
End Function
```

Figure 13.12b

(b) Go to Sheet5 of the workbook. On this worksheet enter the text and values shown in A1:D5 of Figure 13.13. Select C3:D5 and name the cells in column D.

(c) Enter the text and values in A7:D9. The formula in C9 is =C8*0.01/C7. Copy this across to F9.

(d) Enter the text in row 11.

(e) In row 12 the formulas are:
B12: 0
C12: =h0

Copy C12 across to F12.

	A	B	C	D	E	F
1	Tank emptying problem			Version 2		
2						
3	Initial height (m)		h0	1		
4	Time increment		incr	0.1		
5	Iterations		iter	50		
6						
7	Tank diameter (m)		1	1	1	1
8	Pipe diameter (cm)		1	2	5	10
9	Ratio of diameters		0.01	0.02	0.05	0.1
10						
11		Time (sec)		Height (m)		
12		0	1.000	1.000	1.000	1.000
13		5	0.998	0.991	0.945	0.791
14		10	0.996	0.982	0.892	0.606
15		15	0.993	0.974	0.841	0.446
16		20	0.991	0.965	0.791	0.311
17		25	0.989	0.956	0.742	0.199
18		30	0.987	0.948	0.696	0.113
19		35	0.985	0.939	0.650	0.051
20		40	0.982	0.930	0.606	0.013
21		45	0.980	0.922	0.564	0.000
22		50	0.978	0.913	0.523	0.000

Figure 13.13

(f) In row 13 enter the formulas
B13: =B12+(incr*iter)
C13: =NewHeight(C12, incr, C$9, iter)

Copy C13 across to F13. Copy C13:F13 down to row 31. Save the workbook.

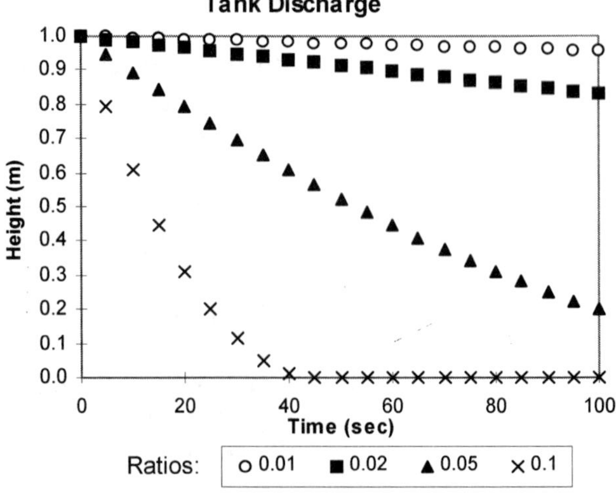

Figure 13.14

We now have a fully operational worksheet. You may wish to add a chart as in Figure 13.14. We may observe the effect of four tank/pipe diameter ratios and, by changing the value in D5, we can observe any time range we wish. You may find (with *iter* = 1) that the results in Version 2 differ in the third decimal place from those in Version 1 for certain sets of conditions. This happens because VBA and the worksheet store values differently and hence have different truncation errors.

Problems

1.* Dynamic systems are often described by a second-order differential equation of the form:

$$\frac{d^2x}{dt^2} = f\left(t, x, \frac{dx}{dt}\right)$$

By letting $v = dx/dt$, this may be transformed into the two simultaneous first-order differential equations:

$$\frac{dv}{dt} = f(t, x, v) \qquad \text{and} \qquad \frac{dx}{dt} = v$$

(a) Show that the Runge–Kutta formulas for these equations may be written as

$$x_{n+1} = x_n + hv_n + \frac{h}{6}(q_1 + q_2 + q_3)$$

$$v_{n+1} = v_n + \frac{1}{6}(q_1 + 2q_2 + 2q_3 + q_4)$$

where:

$$q_1 = hf(t_n, x_n, v_n)$$
$$q_2 = hf(t_n + \tfrac{1}{2}h, x_n + \tfrac{1}{2}hv_n, v_n + \tfrac{1}{2}q_1)$$
$$q_3 = hf(t_n + \tfrac{1}{2}h, x_n + \tfrac{1}{2}hv_n + \tfrac{1}{4}hq_1, v_n + \tfrac{1}{2}q_2)$$
$$q_4 = hf(t_n + h, x_n + hv_n + \tfrac{1}{2}hq_2, v_n + q_3)$$

(b) Use this equation set to study the behaviour of the system depicted in Figure 13.15. The equation of motion for this system is

$$m\frac{d^2x}{dt^2} = -c(x^2 - 1)\frac{dx}{dt} - kx$$

where x is the displacement from equilibrium, m is the mass, c is the damping constant and k is the spring constant. How does your model behave with large and small values of c?

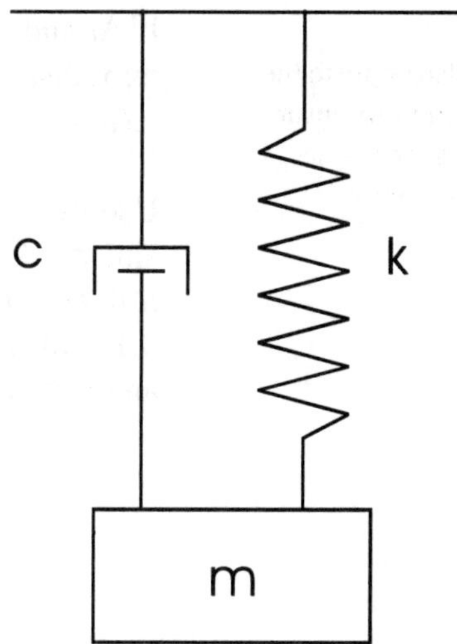

Figure 13.15

(c) The special case where $m = 1$ and $k = 1$ was found by Van der Pol to describe certain electric circuits. Model this system with $x(0) = 0.75$, $v(0) = 0$ and $c = 2$. Graph your results.

2.* Consider a large metal plate of thickness L having a certain temperature profile $u(x,t)$ where x is the distance from one face of the plate and t is the time. The task is to model the plate as it comes to thermal equilibrium given $u(x,0)$.

One-dimensional heat flow is governed by the parabolic partial differential equation:

$$\frac{\partial^2 u}{\partial x^2} = \frac{c\rho}{k}\frac{\partial u}{\partial t}$$

Substitution of the forward-difference equations

$$\frac{\partial^2 u}{\partial x^2} = \frac{u(j+1) - 2u(j,i) + u(j,i-1)}{(\Delta x)^2}$$

and

$$\frac{\partial u}{\partial t} = \frac{u(j+1,i) - u(j,i)}{\Delta t}$$

into the heat-flow equation yields:

$$u(j+1,i) = \frac{k\Delta t}{c\rho(\Delta x)^2}\left(u(j,i+1) + u(j,i-1)\right) + \left(1 - \frac{2k\Delta t}{c\rho(\Delta x)^2}\right)u(j,i)$$

‡ Other conditions also apply to the size of these two terms to ensure stability and convergence – see a numerical analysis textbook.

If Δt and Δx are chosen‡ such that $k\Delta t/c\rho(\Delta x)2$ = ½ the last term disappears, resulting in the simple equation $u(j+1, i) = ½(u(j, i+1) + u(j, i-1))$.

Use this result to model a steel plate for which $u(x,0) = 100 \sin(\pi x/L)$. Let $L = 2$ cm; the boundary conditions are therefore $u(0, t) = 0$ and $u(2, t) = 0$. For steel $c = 0.11$ cal g^{-1} °C^{-1}, $k = 0.13$ cal sec^{-1} cm^{-1} °C^{-1} and $\rho = 7.8$ g cm^{-3}. Hint: divide L into an even number of intervals and think about the symmetry.

14
Statistics for Experimenters

Concepts

Microsoft Excel is a powerful tool for statistical analysis. In this chapter we look at a very small subset of these tools. The main focus is on the treatment of variability associated with data measurements. Some of the functions used in this chapter are:

There has been some criticism of the algorithms used by Excel for its statistical functions. In extreme situations these can lead to errors. For example, while STDEV correctly returns 1 for the data set {1,2,3}, it incorrectly returns 0 for {90000001, 90000002, 90000003}. However, the statistician Neil Cox has concluded (www. agreserach. co.nz/science/statistics/ exceluse.htm) that 'the errors are very unlikely to arise in typical scientific data analysis'.

AVERAGE	Calculates the arithmetic mean of the values in a data set.
DEVSQ	Calculates the sum of the squares of the deviations of the values from their mean.
FREQUENCY	Calculates how often values in a data set occur within a range of values in a bin.
STDEV	Calculates the standard deviation of the values in a data set.
TDIST	Calculates the probability for Student's *t*-distribution.
TINV	Calculates the *t*-value of Student's *t*-distribution.
TTEST	Calculates the probability associated with Student's *t*-test.

We shall also use some of the Data Analysis tools provided by the Analysis ToolPak.

Exercise 1: Descriptive Statistics

An experimenter has collected 100 measurements and wishes to know some statistics of the set; for example, the average, the sum, etc. To save the task of entering 100 numbers we will have Excel generate some random numbers. The RAND or RANDBETWEEN functions are not appropriate since they generate uniform distributions of values. So we will use the Random Number generator tool. To simulate the results from an experiment, we will request random numbers with a mean (average) of 10 and a standard deviation of 0.5.

(a) Open a new workbook. In A1 of Sheet1 enter the label **data**. Use the command Tools|Data Analysis and select the item *Random Number Generation*. Complete the dialog box as shown in Figure 14.1.

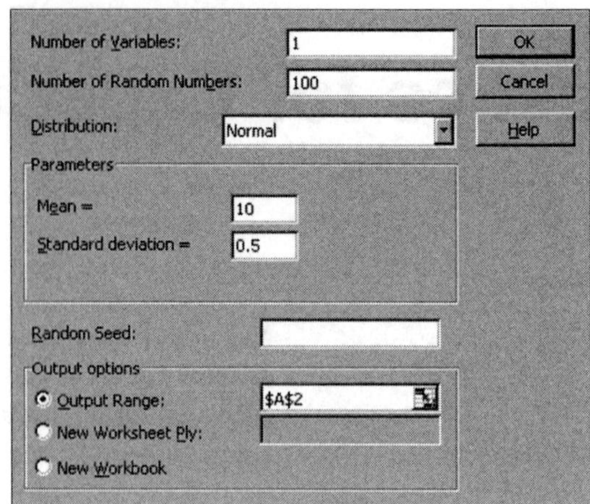

Figure 14.1

(b) To generate the statistics quickly, we will use another Data Analysis tool, namely *Descriptive Statistics*. Complete the dialog box as shown in Figure 14.2. Your worksheet will resemble columns A to D of Figure 14.3 but with slightly different values since we are working with random numbers.

It is important to note the Descriptive Statistics tool generates *values* not formulas. This means that its results are static and do not change as the source data is changed.

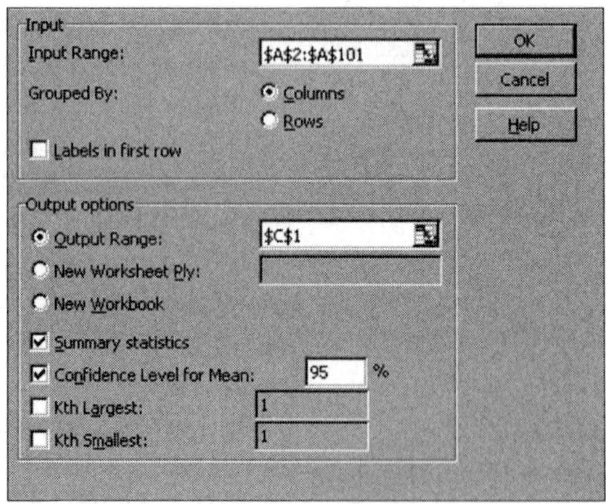

Figure 14.2

(c) There is a worksheet function corresponding to all but two of the statistics generated by the Data Analysis tool. The functions are shown in column F of Figure 14.3. The Confidence Level value returned by the tool differs from that returned by the CONFIDENCE function. We return to this later.

(d) Save the workbook as CHAP14.XLS.

	A	B	C	D	E	F
1	data		*data*		*formulas*	
2	10.51502					
3	10.92899		Mean	10.01859	10.0185879	=AVERAGE(A2:A101)
4	9.718102		Standard Error	0.05162	0.05162	=E7/SQRT(E15)
5	10.21429		Median	9.98280	9.98280	=MEDIAN(A2:A101)
6	10.72602		Mode	#N/A	#N/A	=MODE(A2:A101)
7	9.657998		Standard Deviation	0.51619	0.51619	=STDEV(A2:A101)
8	10.09678		Sample Variance	0.26645	0.26645	=VAR(A2:A101)
9	9.897486		Kurtosis	-0.13466	-0.13466	=KURT(A2:A101)
10	10.40486		Skewness	0.04057	0.04057	=SKEW(A2:A101)
11	9.742392		Range	2.59166	2.59166	=MAX(A2:A101)-MIN(A2-A101)
12	9.884639		Minimum	8.70692	8.70692	=MIN(A2:A101)
13	9.292777		Maximum	11.29858	11.29858	=MAX(A2:A101)
14	10.5603		Sum	1001.85879	1001.85879	=SUM(A2:A101)
15	9.768178		Count	100	100	=COUNT(A2:A101)
16	9.776705		Confidence Level (95.0%)	0.10242	0.00324	=CONFIDENCE(0.95,E7,E15)

Figure 14.3

Exercise 2: Frequency Distribution

Frequently an experimenter wishes to compare the distribution of experimental data with the normal Gaussian distribution. We will use the data generated in the previous exercise rounded to two decimal places.

(a) On Sheet2 of CHAP14.XLS, enter the text values shown in Figure 14.4. In A2 enter =ROUND(Sheet1!A2,2). The pointing method works well here. Copy the formula down to row 101. Name the range A2:A101 as data.

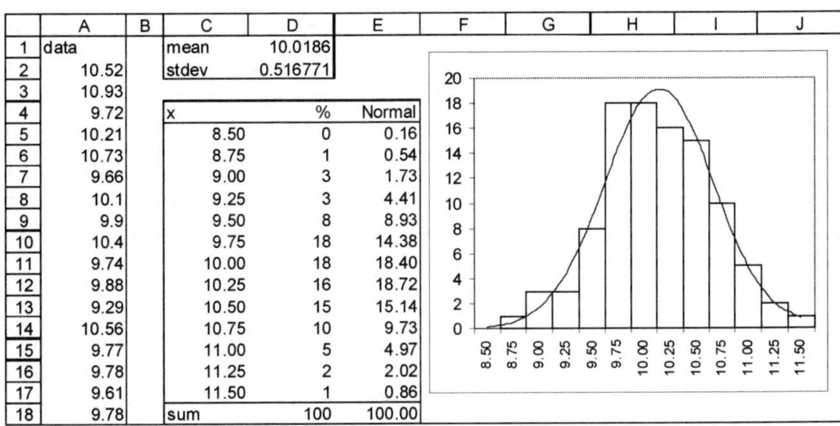

	A	B	C	D	E
1	data		mean	10.0186	
2	10.52		stdev	0.516771	
3	10.93				
4	9.72		x	%	Normal
5	10.21		8.50	0	0.16
6	10.73		8.75	1	0.54
7	9.66		9.00	3	1.73
8	10.1		9.25	3	4.41
9	9.9		9.50	8	8.93
10	10.4		9.75	18	14.38
11	9.74		10.00	18	18.40
12	9.88		10.25	16	18.72
13	9.29		10.50	15	15.14
14	10.56		10.75	10	9.73
15	9.77		11.00	5	4.97
16	9.78		11.25	2	2.02
17	9.61		11.50	1	0.86
18	9.78		sum	100	100.00

Figure 14.4

(b) Compute the mean and the standard deviations of the data using the formulas =AVERAGE(data) and =STDEV(data) in D1 and D2, respectively.

(c) Enter the values in C5:C17. These will serve as the *bin* values or value intervals for the frequency formula.

(d) With D5:D17 selected, enter =FREQUENCY(data, C5:C17) and use [Shift] + [Ctrl] + [Enter] to complete the formula.

(e) Compute the sum of the frequencies with =SUM(D5:D17) in D18. Remember, we are using random numbers, your data will be exactly the same as in the figure.

(f) To compute the expected frequencies use these formulas:
E5: =$D$18 * NORMDIST(C5, mean, stdev, TRUE)
E6: =$D$18 * (NORMDIST(C6,mean,stdev,TRUE) – NORMDIST(C5, mean, stdev, TRUE))

Copy this down to row 17, and modify the formula in E17 to:
E17: =$D$18 * (1 – NORMDIST(C16, mean, stdev, TRUE))

(g) Compute the sum of the expected frequencies with the formula =SUM(E5:E17) in E18. The result should be close to that in D18.

In E5 we compute the cumulative probability for an x value of 8.5 or less. In E6 to E16 we compute the expected frequency for a range of values given by two consecutive x values. In E17 we find the expected probability for values of 11.5 and over.

(h) Make a Line chart of the data in C5:E17. Microsoft Excel will mistakenly think column C contains a data series rather than being the x-category values. You will need to remove the x series by opening the *Series Tab* in Step 2 of Chart Wizard. At the same time, set C5:C17 as the *Category (X) axis labels* for the other series.

(i) Right click on the actual frequency data series. Set the *Chart type* for this series to be *Column*. In the *Options* tab of the *Format Data Series* dialog, set the *Gap width* to 0. Format the *Normal* data series as a smooth line.

(j) Save the workbook.

Exercise 3: The Confidence Limits

Measurements are often repeated a number of times and the results are summarized by an average which statisticians call the *mean*. In addition to reporting the mean value it is often necessary to give an indication of the spread of the measurements. The most commonly used measure of spread is the *standard deviation*. Statisticians speak of *population* and *sample* standard deviations. In theory, a measurement could be repeated an infinite number of times. Our actual measurements are a subset of this hypothetical set, so we use the sample standard deviation. Hence the appropriate Microsoft Excel function is STDEV rather than STDEVP. The data in column A of Figure 14.4 might be reported as *The value of x was found to be 10.019 ± 0.0517 (n = 100)*. In many cases it would be appropriate to use only two decimal places since that was the precision of the raw data.

Our multiple measurements result in a mean value ($\bar{x}$) while the population (true) average is μ, and we would like some way of expressing how close to the real value we think our result is. If I wish to say *I have reason to believe with 90% confidence that μ = 2.45 ± 0.08 (n = 5)*, then the value 90% is referred to as the *confidence level* and 2.45 ± 0.08 is referred to as the *90% confidence limit* (or *interval*). The confidence limits are computed using the Student *t*-statistic:

$$\text{confidence limits for } \mu = \bar{x} \pm \frac{ts}{\sqrt{n}}$$

where t is the Student t-value, s the standard deviation and n the number of measurements. The value of s is found using the STDEV function. We find t using the TINV function which has the syntax TINV(*probability, degrees of freedom*). The probability (α) is (1 – the confidence level expressed as a percentage). For repeated measurements of the same object, the degrees of freedom (f) is given by $n - 1$.

We begin by finding the mean and confidence limits of a set of seven measurements. At the end of the exercise we will make the worksheet more flexible.

(a) On Sheet3 of CHAP14.XLS enter the text shown in columns A to D of Figure 14.5. Ignore columns F to I temporarily. Enter the values in column A.

	A	B	C	D	E	F	G	H	I
1	Mean, standard error and confidence limits								
2	data		mean	1.9543		data		mean	1.9760
3	1.93		stdev	0.0408		1.94		stdev	0.0369
4	1.92		n	7		1.99		n	10
5	2.02		stderror	0.0154		1.98		stderror	0.0117
6	1.97		conf level	95.0%		2.03		conf level	95.0%
7	1.98		t	2.4469		2.03		t	2.2622
8	1.96		conf limits	0.0377		1.96		conf limits	0.0264
9	1.90					1.95			
10			rounded values			1.96		rounded values	
11			mean	1.95		1.92		mean	1.98
12			conf limits	± 0.04		2.00		conf limits	± 0.03

Figure 14.5

(b) The formulas in column D are:

D2:	=AVERAGE(A3:A9)	Mean
D3:	=STDEV(A3:A9)	Standard deviation
D4:	=COUNT(A3:A9)	Number of measurements
D5:	=D3/SQRT(D4)	Standard error of the mean
D6:	=95%	Required level[1]
D7:	=TINV(1–D6, D4–1)	Student's t-statistic (for $\alpha = 0.5, f = 6$)
D8:	=D7*D5	Confidence limit
D11:	=D2	Mean, formatted to two places
D12:	=D8	Confidence limits (formatted[2])

(1) The value may be entered by typing 95%, or by typing 0.95 and formatting with the Percentage tool.
(2) D12 is given a custom format of ±0.00. The symbol is produced with Alt + 0177.

Some journals would require the author to report these values has having a mean of 1.95 and a standard error of 0.015 with $n = 7$. From this information the reader can compute the confidence limits for any required probability level using the formula *confidence limits* = $\pm t \times$ *standard error*. Our worksheet allows the same. You may change the confidence level value in D6 to 95.5, say, to find new confidence limits.

In the days before computers (BC) one had to look up the t values in a table. If you wish to compare the results from TINV with a printed table you should note that TINV gives the two-tailed statistic.

If we wished we could simply compute the confidence limit with one formula:

```
=STDEV(A3:A9) * TINV(1 – D6, COUNT(A3:A9) – 1) /
SQRT(COUNT(A3:A9)).
```

Complex formulas such as this, however, are error-prone.

Our worksheet would be useful for any experiment in which a measurement is repeated seven times or less. We can test this and at the same time double-check our worksheet.

(c) Use the Descriptive Statistics tool with the data in A3:A9. Do the values it reports for the mean, standard error, standard deviation and confidence limits agree with your worksheet? Erase the values in A3:A9 and enter three new values. Use the Descriptive Statistics tool again (you will recall that its values are static and must be recomputed manually when the data changes) and check for agreement.

Our worksheet will not give correct results with more than seven data items unless we make appropriate changes to all the formulas in column D that reference A3:A9. We can, however, make the worksheet flexible.

(d) Copy A2:D12 to F2. Modify the formulas in column I to read:
I2: =AVERAGE(F:F)
I3: =STDEV(F:F)
I4: =COUNT(F:F)

The range references to F:F may be interpreted as F1:F65536. This means the worksheet will give the correct result no matter how many values are entered. The empty cell in F1 and the text in D2 have no effect. Enter the values shown in F1:F12 and use Descriptive Statistics to validate your worksheet results.

In Exercise 1 we saw that the CONFIDENCE function result does not agree with the results reported by the Descriptive Statistic tool. This function always uses a t-value for an infinite value of f, the degrees of freedom. Its results may be acceptable when n is very large. The other case where this gives an acceptable value is when, from experience, we know that the standard deviation (s) for the n measurement is always close to the standard deviation (σ) for a large number of measurements.

Exercise 4: Experimental and Expected Mean

A series of measurements may be made on a sample where there is an expected result. A chemist may analyse a sample thought to be compound X and compare the results with the known composition of X to determine if the sample is pure X. An engineer may measure the thickness of a metal plate and compare the results with the known thickness to test a new measuring device. Statisticians speak of hypothesis testing in these cases. For the chemist, we have the null hypothesis H_0: *The sample is compound X.* For the engineer, we have H_0: *This new instrument is suitable for the task.* Both may be stated as H_0: *This measured average ($\overline{x}$) is the same as the expected average (μ).* There is the alternate hypothesis H_1: *This measured average ($\overline{x}$) differs from the expected average (μ).* We compute an experimental *t*-statistic and compare it with the critical value‡ for a given confidence level. If the experimental *t* does not exceed the critical *t*, we dismiss the alternative hypothesis.

‡ Statistical textbooks often talk of *t(calculated)* and *t(table)*. The terms *t(experimental)* and *t(critical)* are used here to avoid confusion since we 'calculate' both values.

The experimental *t* is computed using

$$t_{experimental} = \frac{|\overline{x} - \mu|}{s_{\overline{x}}} = \frac{|\overline{x} - \mu|}{s_x / \sqrt{n}}$$

where $s_{\overline{x}}$ is the standard deviation of the mean and s_x is the standard deviation in the *n* measurements.

To calibrate a packing machine, an engineer has made a series of measurements of the raisin content in boxes of breakfast cereal. The required value is 33%. To test the results we will construct a worksheet similar to that in Figure 14.6.

	A	B	C	D	E	F	G
1	Experimental and Expected Mean						
2							
3	data		Method 1			Method 2	
4	30.3		expected mean	33		expected mean	33
5	34.7		experiment mean	35.34		experiment mean	35.34
6	40.0		stdev	4.238		stdev	4.238
7	36.1		n	7		n	7
8	41.3		t expt	1.46		t expt	1.46
9	34.5		prob (required)	95%		alpha (required)	0.05
10	30.5		t(α, df)	2.45		prob (t, df, 2)	0.19
11			Null hypothesis			Null hypothesis	
12							
13							

Figure 14.6

(a) On Sheet4, enter the text shown A1:C11 of Figure 14.6. For the time being, ignore the entries in columns F and G. Enter the experimental values in column A. Select A3:A13 and name A4:A13 as data. This will allow the worksheet to be used with up to 10 measurements although we have only seven.

(b) The entries in column D are:
D4: 33 Required mean

D5: =AVERAGE(data) Calculated mean
D6: =STDEV(data) Standard deviation
D7: =COUNT(data) Number of measurements
D8: =ABS(D4–D5)/(D6/SQRT(D7)) Experimental *t*-value
D9: 95% Required level of confidence
D10: =TINV(1 – D9, D7 – 1) Critical *t*-value
D11: =IF(D8>D10,"Alternate hypothesis","Null hypothesis")

The entry in D11 is centred across D11 and D12 using the Merge and Center tool.

In this case the experimental *t*-value (1.46) does not exceed the critical value (2.45) so the null hypothesis is considered accepted. With 95% certainty, we may say there is no statistical difference between the found and the expected mean.

The alternative approach to this problem is to compute the probability that the mean value is statistically different from the expected value. We use TDIST to compute the *p*-value from the data and compare this to our required α-value which is generally 0.05 or 5%. The syntax for TDIST is TDIST(*x*, *df*, *tails*), where *x* is the value to be evaluated, *df* is the degrees of freedom and *tails* has a value of 1 for a one-tail distribution or 2 for a two-tailed.

(c) Copy C3:D11 to F3. Edit the text in F3 and F10. Edit the text in F9 and F10 and makes these changes:
G9: =1 – D9 The required α-value
G10: =TDIST(G8,G7 – 1,2) The computed *p*-value
G11: =IF(G9<G10,"Null hypothesis","Alternate hypothesis")

We accept the null hypothesis (that the two means are statistically the same) since the calculated *p* (0.19) is greater than the stipulated α-value (0.05). We may interpret these results as saying that if the null hypothesis is true there is a 19% probability that seven boxes taken at random will show a difference of 2.34 (=35.34 – 33.00) from the expected mean of 33. Note that we are saying that the difference between the found and expected means could occur by random errors. We are not necessarily saying this is an acceptable situation. The engineer may accept the accuracy of the machine but may decide to improve its precision in order to decrease the spread of the values.

In this problem we used a two-tailed *t*-value (the last argument in the formula) since we were concerned with both positive and negative differences from the expected mean. Consider another scenario. The packing machine fills, on average, 50 boxes a minute. After modification, 10 trials were made and the average filling rate was found to be 54.5 boxes/min with a standard deviation of 4.3. Has there been a statistically significant improvement in the machine? From these values t computes to 3.31 and a one-tailed p-value is 0.0035. So at the 5% level there has been no improvement, since p is less than α.

Exercise 5: Pooled Standard Deviation

This exercise shows one way to treat repeated measurements on different samples. We introduce the concept of the *pooled standard deviation* and the function DEVSQ. A biologist has measured the mercury content of seven fish taken from Lake Erie and obtained the results[‡] shown in Figure 14.7.

‡ Data taken from D. A. Skoog and M. W. West, *Analytical Chemistry*, 2nd ed., p. 40, New York: Holt, Reinhart and Winston, Inc., 1974.

	A	B	C	D	E	F	G	H	I	J
1	Sample	Replicates			Results				Mean	SSD
2	1	3	1.80	1.58	1.64				1.6733	0.02587
3	2	4	0.96	0.98	1.02	1.10			1.0150	0.01150
4	3	2	3.13	3.35					3.2400	0.02420
5	4	6	2.06	1.93	2.12	2.16	1.89	1.95	2.0183	0.06108
6	5	4	0.57	0.58	0.54	0.59			0.5700	0.00140
7	6	5	2.35	2.44	2.70	2.48	2.44		2.4820	0.06848
8	7	4	1.11	1.15	1.22	1.04			1.1300	0.01700
9	7	28							ss ->	0.20953
10										
11			mean of all measurements				1.67			
12			pooled standard deviation				0.10			

Figure 14.7

The pooled standard deviation is computed using the formula:

$$s_{pooled} = \sqrt{\frac{\sum SSD_i}{\sum r_i - n}} \quad \text{or} \quad s_{pooled} = \sqrt{\frac{\sum SSD_i}{\sum (r_i - 1)}}$$

where SSD_i is the sum of the squares of the deviations from the mean for the ith sample and r_i is the number of repeated measurements on the i-th sample. The degrees of freedom are given in the divisors in these two equivalent formulas. How can we set up a general worksheet to handle measurements of this type?

(a) On Sheet5 of CHAP14.XLS enter the text shown in the figure. Enter the values shown in A2:A8 and C2:H8.

(b) The number of samples n is found in D9 with the formula =COUNT(A2:A9). The number of repeated measurements for the first sample (r_1) is found in B2 with =COUNT(C3:H3) and this is copied down to B4. The total number of measurements ($\sum r_i$) is found in B9 with =SUM(B2:B9).

(c) The mean for the first sample is found in I2 with =AVERAGE(C2:H2). The sum of the squares of the deviations from the mean for the first sample (SSD_1) is found in J2 with =DEVSQ(C2:H2). These formulas are copied down to row 8. The sum of the SSD values is computed in J9 with =SUM(J2:J8).

(d) The mean for all measurements is given in G11 by =AVERAGE(C2:H8). The formula in G12 for pooled standard deviation is =SQRT(J9/(B9 – A9)).

Exercise 6: Comparing Paired Arrays

In this exercise we compare the mean from two sets of measurements made on a set of samples. Perhaps set A are the measurements using one technique while set B were obtained from another. As in Exercise 4, we compute a standard deviation and use it to find a *t*-value. We compare the found *t*-value with the critical value computed for a specified α-value and the appropriate degrees of freedom. As before, we accept the null hypothesis if the experimental *t*-value is less than the critical *t*-value.

For these circumstances (two measurements on several different samples) the *t*-value is computed using:

$$t_{expt} = \frac{\bar{d}}{s_d}\sqrt{n}$$

where:
n is the number of paired measurements
$\bar{d}$ is the average of the differences between the pairs and

$$s_d = \sqrt{\frac{\sum(d_i - \bar{d})^2}{n-1}}$$

In Exercise 4 we saw two methods to compare a measured mean with an expected value. We could call these the *t method* and the *p method.* We can also use a probability method for paired arrays of data. We will not show the mathematical equations associated with this, but will use the TTEST function which has the syntax: TTEST(*array1, array2, tails, type*) where *tails* has the same meaning as before, and *type* is given a value of 1 for paired arrays.

	A	B	C	D	E	F	G	H	I	J	K
1	Paired sets										
2		Data				Calculations			t-Test: Paired Two Sample for Means		
3	Sample	A	B	diff		mean diff	0.16				
4	1	1.11	0.97	0.14		n	6			*A*	*B*
5	2	3.77	4.33	-0.56		SSD	1.029		Mean	3.165	3.003
6	3	5.94	5.35	0.59		st dev	0.454		Variance	3.602	3.191
7	4	2.90	2.30	0.60		t expt	0.873		Observations	6	6
8	5	1.04	1.19	-0.15		alpha	0.05		Pearson Correlation	0.971	
9	6	4.23	3.88	0.35		t critical	2.571		Hypoth Mean Diff	0	
10						decision	same		df	5	
11									t Statistic	0.873	
12									P(T<=t) one-tail	0.211	
13						p approach			t Critical one-tail	2.015	
14						p from t-test	0.423		P(T<=t) two-tail	0.423	
15									t Critical two-tail	2.571	

Figure 14.8

(a) On Sheet6 of CHAP14.XLS, enter the text values shown in columns A:G of Figure 14.8. For now, ignore columns I:K. Enter the values shown in A4:C9.

(b) Enter the formula =B4–C4 in D4 and copy it down to row 9.

(c) The formulas in column G are:

G3: =AVERAGE(D4:D9) Computes $\bar{d}$
G4: =COUNT(A4:A9) Computes n
G5: =DEVSQ(D4:D9) Computes $\sum(d_i - \bar{d})^2$
G6: =SQRT(G5/(G4–1)) Computes s_d
G7: =G3*SQRT(G4)/G6 Compute *t*(*experimental*)
G8: 0.05 The required α-value
G9: =TINV(G8, G4–1) Computes *t*(*critical*)
G10: =IF(G7<G9,"same","not same")
G14: =TTEST(B4:B9,C4:C9,2,1) Computes the *p*-value

We are led to the conclusion that the two methods give the same mean (with an α-value of 0.05) since (i) *t*(*experimental*) is less than *t*(*critical*) and (ii) the *p*-value computed by TTEST is greater than the alpha value or 0.05.

To round off this exercise, we use the *t-TEST: Paired Two Sample for Means* tool from the Data Analysis tool. This is left as an exercise for the reader. Note that you should set the *Hypothesized mean difference* to 0 and the *alpha* value to 0.05 when completing the tool's dialog box. The results are shown in the figure. As expected, the results agree with our own calculations. The *t*(*experimental*) values in G9 and J15 are the same, as are the *p*-values in G14 and J14. These serve as useful checks but recall that the results from the tool are static whereas our calculations will be updated if new experimental array values are entered.

Exercise 7: Comparing Repeated Measurements

In the previous exercise each sample was measured once by each of two techniques. In this exercise the same sample is measured repeatedly by two techniques. Our task is the same, to determine if the mean of the two sets of measurements is the same. Once again, we have two statistical methods we could use: the *t* and the *p* methods. For the former we compute a pooled standard deviation using the formula‡:

$$s_p = \sqrt{\frac{\sum_{set1}(x_i - \bar{x}_i)^2 + \sum_{set2}(x_j - \bar{x}_j)^2}{n_1 + n_2 - 1}} = \sqrt{\frac{s_1^2(n_1 - 1) + s_2^2(n_2 - 1)}{n_1 + n_2 - 2}}$$

‡ When the two data sets are of equal size, this reduces to $s_p = \sqrt{(s_1^2 + s_2^2)/2}$.

from this we compute *t(experimental)* and compare it with *t(critical)*. The experimental *t*-value is found using:

$$t_{experimental} = \frac{\bar{x}_1 - \bar{x}_2}{s_p}\sqrt{\frac{n_1 n_2}{n_1 + n_2}} = \frac{\bar{x}_1 - \bar{x}_2}{s_p \sum(1/n_1 + 1/n_2)}$$

For the *p* method we will again use the Microsoft Excel functions TDIST or TTEST to find a probability value which we will compare to the required α-value. We will also use the Data Analysis tool *t-Test: Two Sample Assuming Equal Variance* to check our results.

(a) On Sheet7 of CHAP14.XLS enter the text shown in A1:D19 of Figure 14.9. Enter the experimental values in columns A and B. Select A4:B19 and use the Insert|Name command to name A1:A19 as A and B1:B19 as B. This will allow the worksheet to be used with up to 15 data points.

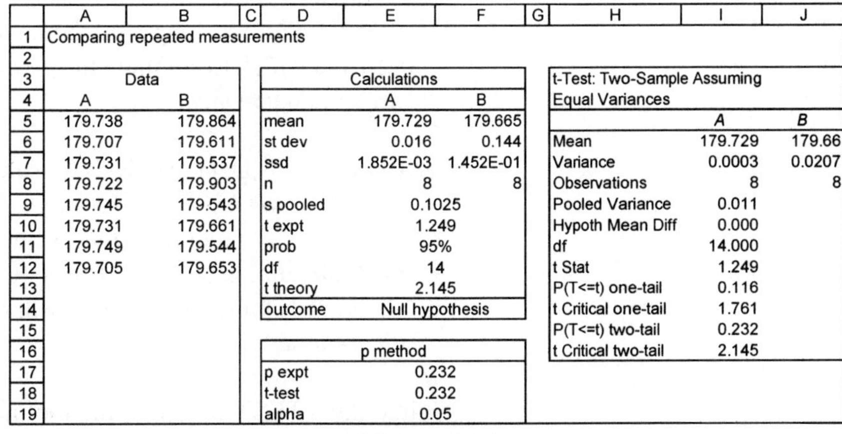

	A	B	C	D	E	F	G	H	I	J
1	Comparing repeated measurements									
2										
3	Data				Calculations			t-Test: Two-Sample Assuming		
4	A	B			A	B		Equal Variances		
5	179.738	179.864		mean	179.729	179.665			*A*	*B*
6	179.707	179.611		st dev	0.016	0.144		Mean	179.729	179.66
7	179.731	179.537		ssd	1.852E-03	1.452E-01		Variance	0.0003	0.0207
8	179.722	179.903		n	8	8		Observations	8	8
9	179.745	179.543		s pooled	0.1025			Pooled Variance	0.011	
10	179.731	179.661		t expt	1.249			Hypoth Mean Diff	0.000	
11	179.749	179.544		prob	95%			df	14.000	
12	179.705	179.653		df	14			t Stat	1.249	
13				t theory	2.145			P(T<=t) one-tail	0.116	
14				outcome	Null hypothesis			t Critical one-tail	1.761	
15								P(T<=t) two-tail	0.232	
16					p method			t Critical two-tail	2.145	
17				p expt	0.232					
18				t-test	0.232					
19				alpha	0.05					

Figure 14.9

(b) The formulas in columns E and F are:
E5: =AVERAGE(A)
E6: =STDEV(A)
E7: =DEVSQ(A)
E8: =COUNT(A)
E9: =SQRT((E7+F7)/(E8+F8–2))
E10: =(ABS(E5 – F5)/E9) * SQRT((E8*F8)/(E8+F8))
E11: 95% The required confidence level
E12: =E8 + F8 – 2 The degrees of freedom
E13: =TINV(1–E11,E12)
E14: =IF(E10>E13,"Alternative hypothesis","Null hypothesis")
F5: =AVERAGE(B)
F6: =STDEV(B)
F7: =DEVSQ(B)
F8: =COUNT(B)

Comparing the *t*(*experimental*) value of 1.249 in E10 with the *t*(*critical*) value of 2.145 in E13, we accept the null hypothesis that the two means are statistically the same.

(c) For the *p* method, the formula in E17 is =TDIST(E10,E12,2) and in E18 it is =TTEST(A, B, 2, 2) for a two-tailed test with sets having equal population variances. In E19 we use =1–E11 to compute the required alpha.

The results here lead to the same conclusion: that the null hypothesis cannot be dismissed.

You may wonder why we used two formulas for the *p* method. The simple answer is that TTEST is only of use when the two arrays are of equal size. The longer method, which involves computing a *t*-value from which to compute the *p*-value, is applicable when the sets are of unequal size.

(d) Using Tools|Data Analysis..., select the tool *t-Test: Two-Sample Assuming Equal Variance*. Set the *Hypothetical mean difference* to 0 and the *alpha* value to 0.05 when completing the tool's dialog box. Use H3 as the *Output range*. The two *t*-statistics from the tool agree with our calculations and so do the *p*-values.

Unlike the TTEST function, the tool may be used with arrays of unequal size. We have been speaking of testing the hypothesis that there is no difference in the mean of two data sets. We could rephrase this to: testing that the means differed by zero. Can we test if the means differ by a non-zero amount? Yes, by entering a value in the *Hypothetical mean difference* box. If we wish to do a similar test with formulas, the *t*(*experimental*) value must be computed using:

$$t_{experimental} = \frac{(\bar{x}_1 - \bar{x}_2) - (\mu_1 - \mu_2)}{s_p}\sqrt{\frac{n_1 n_2}{n_1 + n_2}}$$

where $(\mu_1 - \mu_2)$ represents the hypothesized difference in the population means.

Exercise 8: The Calibration Curve Revisited

In Chapter 7 we saw how to chart a calibration curve and add a trendline. We also used the functions SLOPE, INTERCEPT and LINEST to find the slope and intercept of the line of best fit. This line, of course, has uncertainties associated with it. The LINEST function not only gives us the values for the slope and intercept, it also gives the errors associated with them. Let s_b be the standard error (uncertainty) for the intercept b, s_m the standard error for the slope m and s_y the standard error for the estimate of y. If y^* is the measured signal for an unknown, then the value of the unknown is computed using

$$x^* = \frac{y^*(\pm s_y) - b(\pm s_b)}{m(\pm s_m)}$$

In this exercise we make a calibration curve and determine x^* for a measured y^* using the equation above. The function LINEST is used to find the required parameters. We will see how a combination of INDEX and LINEST allows us to generate only those parameters that are necessary for the task. We shall need to recall that errors are combined using $e_3 = \sqrt{e_1^2 + e_2^2}$ and that for multiplication and division, we must work with percentage errors.

(a) On Sheet8 of CHAP14.XLS enter the text shown in Figure 14.10.

(b) Enter the calibration data in A4:B8. Name the columns as x and y, respectively.

(c) Select D4:E8, enter the formula =LINEST(y, x, TRUE, TRUE) and press Ctrl+Shift+Enter to complete the array formula. The entry will appear in the formula bar surrounded by braces {} because it is an array formula.

(d) To see how we may obtain certain parameters from the LINEST function, enter the formulas shown below. These are not array formulas so complete them normally.
B12: =INDEX(LINEST(y, x, TRUE, TRUE), 1,1)
B13: =INDEX(LINEST(y, x, TRUE, TRUE), 1,2)
B14: =INDEX(LINEST(y, x, TRUE, TRUE), 2,1)
B15: =INDEX(LINEST(y, x, TRUE, TRUE), 2,2)
B16: =INDEX(LINEST(y, x, TRUE, TRUE), 3,2)

The first formula returns the LINEST value that would normally be in the first row and first column, i.e. the slope of the line of best fit. Likewise, the second gives us the intercept which is in row 1, column 2, of the LINEST array.

	A	B	C	D	E	F
1	Uncertainty in a Calibration Curve					
2						
3	x	y			Linest	
4	1	2.86	slope	2.288	0.568	intercept
5	2	5.20	error in slope	0.030022	0.099572	error in intercept
6	3	7.40	R^2	0.999484	0.094939	error in estimate of y
7	4	9.60	F statistic	5808	3	degrees of freedom
8	5	12.10	regression ss	52.34944	0.02704	residual sum of squares
9						
10						
11	Index with Linest			value	err	%err
12	m	2.288	y*	6.55		
13	b	0.568	numerator	5.982	0.1376	2.30%
14	sm	0.0300	denominator	2.288	0.0300	1.31%
15	sb	0.0996	x*	2.615	0.0692	2.65%
16	sy	0.0949				

Figure 14.10

(e) Select A12:B16 and use the Insert|Name command to name the cells in B12:B16. This will make it easier to understand the formulas that follow.

(f) For the purpose of the exercise, assume our measured signal had a value of 6.55. Enter this value in D12. Enter the following formulas:

D13: =D12 – b	The numerator $(y^* - b)$
E13: =SQRT(sy^2+sb^2)	The error in the nominator
F13: =E13/D13	The percentage error in the nominator
D14: =m	The denominator m
E14: =sm	The error in the denominator
F14: =E14/D14	The percentage error in the denominator
D15: =D13/D14	The value $x^* = (y^* - b)/m$

E15: =D15*F15	The error in x^*. This will mean nothing until F15 is computed
F15: =SQRT(F13^2 + F14^2)	The percentage error in x^*

When using a spreadsheet (or a calculator) to do such computations, we let it use its full precision. We may wish to format the cell to show a limited number of digits if the spreadsheet is to be displayed to others. We must round off the values when reporting the results. We would report x^* as $2.59_3 \pm 0.07_0$ or $2.59_3 \pm 2._7\%$.

Exercise 9: More on the Calibration Curve

‡ See, for example, P. C. Meier and R. E. Zünd, *Statistical Methods in Analytical Chemistry*, New York: Wiley, 1993.

The statistical analysis in the previous exercise ignores the fact that the estimations of the slope and intercept are interdependent. A full treatment of the alternative approach is beyond the scope of this book. To enable the reader to use Microsoft Excel to perform the calculations that are given in advanced statistics books‡, we shall show without comment the formulas used.

When the advanced treatment is used to compute the upper and lower confidence intervals for the line of best fit, curves as shown in Figure 14.11 result. Note that very poor data was purposely used to get a figure in which the two confidence intervals are visible.

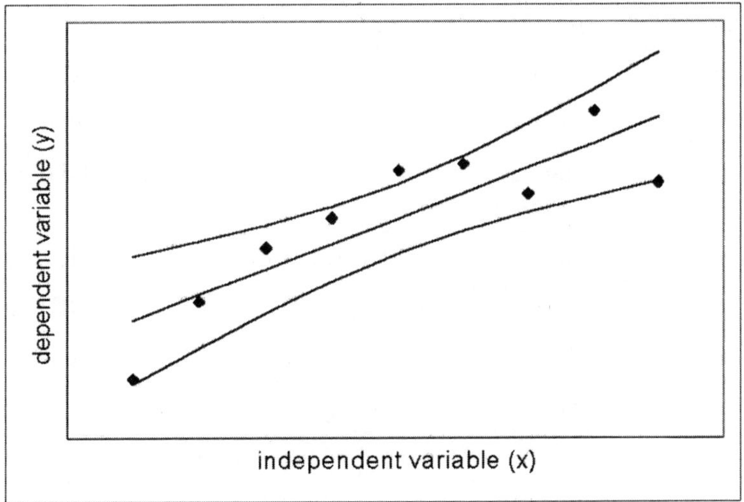

Figure 14.11

Note: Statisticians use the symbol *b* for the slope and the symbol *a* for the intercept in regression analysis. This can be confusing for those of us who use $y = mx + b$ for the equation of a straight line.

The expression for the confidence interval for the computed *Y*-values is:

$$CI(Y) = \pm t(\alpha, df) \cdot s_{res} \cdot \sqrt{\frac{1}{n} + \frac{(x - \bar{x})^2}{S_{xx}}}$$

The confidence interval for the predicted x^*- value is found using one of:

$$CI(x^*) = \pm t(\alpha, df) \cdot \frac{s_{res}}{|m|} \cdot \sqrt{\frac{1}{n} + \frac{1}{k} + \frac{(y^* - \bar{y})^2}{m^2 \cdot S_{xx}}}$$

$$CI(x^*) = \pm t(\alpha, df) \cdot \frac{s_{res}}{|m|} \cdot \sqrt{\frac{1}{n} + \frac{1}{k} + \frac{(x^* - \bar{x})^2}{S_{xx}}}$$

The variables needed to compute these expressions are shown in the table below, together with the appropriate Excel functions.

Symbol	Purpose	Excel
$x_1, x_2, \ldots, x_i$ $\bar{x}$	the independent values used in the calibration, and their average	a range named *x* AVERAGE(x)
$y_1, y_2, \ldots, y_i$ $\bar{y}$	the dependent values used in the calibration, and their average	a range named *y* AVERAGE(y)
n	the number of *x*,*y* pairs	COUNT(X)
df	degrees of freedom = $n - 2$	$n - 2$
m	the slope of the line	INDEX(LINEST(y,x,TRUE, TRUE), 1,1)
b	the intercept of the line	INDEX(LINEST(y,x,TRUE, TRUE), 1,2)
s_{res}	standard deviation in sum of the squares of the residuals	INDEX(LINEST(y,x,TRUE, TRUE), 3,2)
S_{xx}	sum of squares of *x* deviations	DEVSQ(x)
df	degrees of freedom	INDEX(LINEST(y,x,TRUE, TRUE), 4,2)
α	the confidence level, generally 0.05	a value
$t(\alpha, df)$	Student's *t*-value for given α and *df*	TINV(alpha, df)
k	number of repeated y^* measurements	COUNT(range)

We begin by using the calibration data from the previous exercise and computing the confidence levels.

(a) Copy A1:B8 from Sheet8 of CHAP14.XLS to A1 of Sheet9. Name the two ranges x and y. Enter the text in A10:A19 of Figure 14.12. Select A10:B19 and name the cells.

	A	B	C	D	E
1	Confidence levels for line of best fit				
2					
3	x	y	Y	Y+CL	Y-CL
4	1	2.86	2.86	3.09	2.62
5	2	5.20	5.14	5.31	4.98
6	3	7.40	7.43	7.57	7.30
7	4	9.60	9.72	9.89	9.55
8	5	12.10	12.01	12.24	11.77
9					
10	Parameters		x* confidence levels		
11	m	2.288	y*	avg y*	6.55
12	b	0.568	6.55	x*	2.615
13	n	5	6.47	CL x*	0.0851
14	df	3	6.56	CL x*	0.0851
15	Sres	0.095	6.57	%	3.25%
16	Sxx	10	6.60		
17	avgx	3			
18	avgy	7.432			
19	alpha	0.05			

Figure 14.12

(b) The required parameters are found with these formulas:
B11: =INDEX(LINEST(y, x, TRUE, TRUE), 1, 1)
B12: =INDEX(LINEST(y, x, TRUE, TRUE), 1, 2)
B13: =COUNT(x)
B14: =n – 2
B15: =INDEX(LINEST(y, x, TRUE, TRUE), 3, 2)
B16: =DEVSQ(x)
B17: =AVERAGE(x)
B18: =AVERAGE(y)
B19: 0.05

(c) The formulas in C4:E4 to compute the predicted Y- values and the upper and lower confidence levels are:
C4: =A4*m + b
D4: =$C4 + TINV(alpha,df) * Sres * SQRT((1/n+($A4–avgx)^2/Sxx))
E4: =$C4 – TINV(alpha,df) * Sres * SQRT((1/n+($A4–avgx)^2/Sxx))

The mixed cell references in D4's entry permits copying it to E4 and then changing the sign. The cells C4:E4 are copied down to row 8.

Finally, we show how to use compute a predicted x-value from a series of sample measurements.

(d) Enter the text shown in C11:D15 of the figure. Enter the measurements, values in value C12:C16. These represent five duplicated analyses of the same sample.

(e) Average the y^*-values with the formula =AVERAGE(C12:C16) in cell E11. The computed x^*-value in E12 is found with the formula =(E11 – b)/m while the confidence intervals in E13 and E14 are found with:

E13: =TINV(alpha,df) * (Sres/m) * SQRT(1/n + 1/COUNT(C12:C16) + (E11–avgy)^2/(m^2*Sxx))

E14: =TINV(alpha,df) * (Sres/m) * SQRT(1/n + 1/COUNT(E12:E16) + (E12–avgx)^2/Sxx)

We have used two formulas merely to show they are equivalent; some texts use one, some the other. The percentage error (uncertainty) is computed in E15 with =E13/E12 and formatted as a percentage.

You will see that this treatment gives a result that differs somewhat from that obtained in the previous exercise. We would report the x^* values as $2.61_5 \pm 0.08_5$ or $2.61_5 \pm 3._3\%$ at the 95% confidence level.

Problems

1. The distribution of weight of 1000 pills from a certain machine was found to be well described by a Normal curve with a mean of 400 mg with a standard deviation of 50 mg. What fraction of the pills are expected to be in the interval 400 ± 10 mg?

2. An analysis of substance X, thought to be compound Q, gave these results for percentage carbon: 59.09, 59.17, 59.27, 59.13, 59.1, 59.14. The expected result for compound Q is 59.55. What conclusion can be drawn?

   Source of data: F. W. Power, *Analytical Chemistry*, **11**, 6000 (1939). The data has a wide spread by modern standards.

3. The *F*-statistic is another measure used to compare data. As with the *t*-statistic, one computes an *F*-value from the data and compares it to a critical *F*-value. The null hypothesis (no difference in the means) is accepted when the group *F*-value is less than the critical value. The *Anova: Single Factor* Data Analysis tool is one way to do this. The table below represents the results from three testing laboratories working with the same sample. Does the Anova result suggest that the mean values of these results are statistically different?

A	7.88	7.73	7.66			
B	7.70	7.49	7.38	7.70		
C	7.18	7.51	7.14	7.79	7.53	7.68

15
Report Writing

Concepts

In this chapter we learn how to place Microsoft Excel workbook data and charts into a word processor document. There are two very different ways to do this: (i) using copy and paste, or (ii) with Object Linking and Embedding (OLE). We will examine these methods in detail in the exercises. The algorithm below will help you choose the appropriate method.

> Are you sure that the workbook is complete and the report will never need updating?
>
> Yes: Use copy and paste.
> No: Will you always have access to the workbook?
> Yes: Use linking.
> No: Use embedding.

Exercise 1: Copy and Paste

The method described in this Exercise will work with Microsoft Office products such a Word and PowerPoint. They also work with many other applications. If a simple Paste does not give the required result, look in the Paste Special options.

In this exercise we will copy data and a chart from an Excel workbook to a document you are writing with a word processor application such as Microsoft® Word or Corel® WordPerfect. We need a simple workbook with data and a chart. Let us assume we have run an experiment to find the value of a resistor by measuring the currents passing through the resistor when various voltages are applied. Since $I = V/R$, the slope of a plot of I vs V will be $1/R$.

(a) Open a new Excel workbook. Enter the data shown in A1:B11 of Figure 15.1. The cell B11 contains the formula =1000/SLOPE(B4:B9,A4:A9) where the factor of 1000 accounts for the fact that the current was measured in milliamps.

(b) Construct a chart similar to that in Figure 15.1. Insert a trendline without the formula being displayed.

Reminder: In Microsoft Office products right-clicking brings up a shortcut menu that is very convenient for copying and pasting.

(c) Save the workbook as CHAP15.XLS.

(d) Without closing Excel, open your word processor. Put some text into a new document as in Figure 15.2 but without the table or chart.

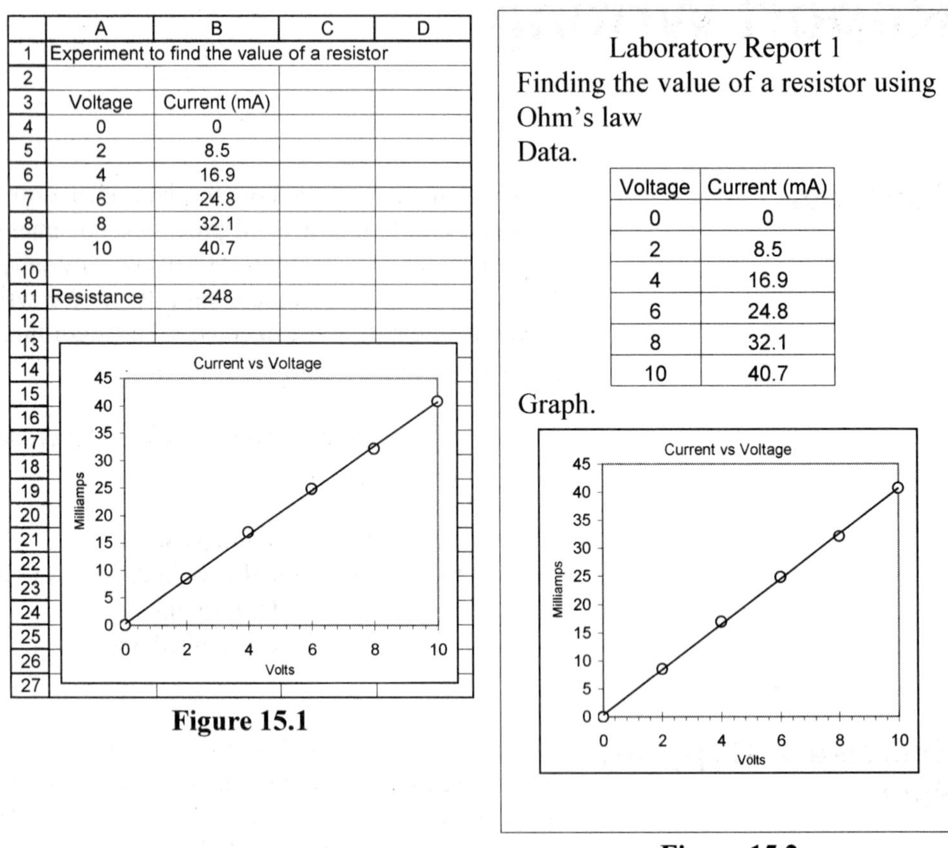

Figure 15.1

Figure 15.2

(e) Make Excel the active application. Select the range A3:B9 in CHAP15.XLS. Either click the COPY button or use Edit|Copy.

(f) Make the word processor active. Move to the line below 'Data' and click the Paste button. The data from the Excel worksheet is inserted into the document as a table. You may need to adjust its position and to add any required borders to the table. Note that the table uses the same font as the worksheet but this may be changed from within the word processor if you so wish.

Should you need the text not in table form, use Edit|Paste Special and stipulate non-formatted text. This will give you the data in tabular columns.

(g) Return to the workbook. Click once on the graph and click the Copy button.

(h) Activate the word processor and move the insertion pointer under the word 'Graph'. From the main menu select Edit|Paste Special. This will bring up a dialog box similar to that in Figure 15.3. Make sure the *Paste* radio button is selected and in the *As* box select *Picture (Enhanced Metafile)*. Click the OK button. The graph is now added to the document as a picture. It may be positioned and sized to suit your needs using the word processor commands.

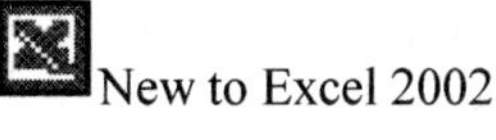

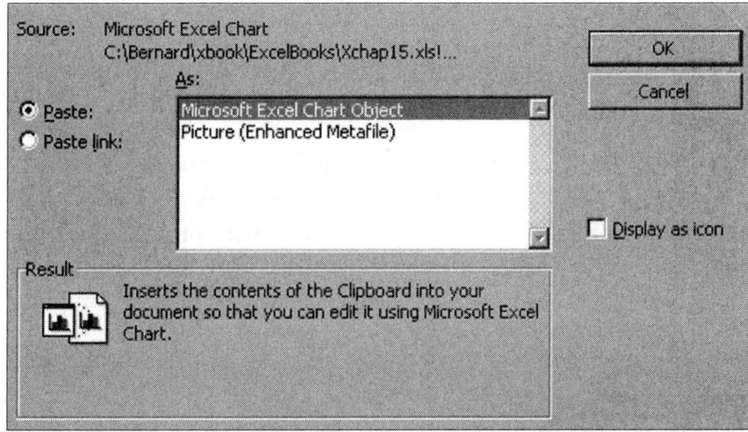

Figure 15.3

The Paste Option in Word 2002 may cause a smart tag to be displayed. This may be used, for example, to change an item copied as a picture to convert to a linked object.

If you need to copy the same Excel object many times, the Windows Clipboard can be inconvenient. Recall that an Excel object remains on the Windows Clipboard only while the object is selected – in the case of a range, only while the 'ant track' is present. The Office Clipboard may be used to hold up to 24 objects and they are all available in the various Office components. The command Edit|Office Clipboard may be used to display the Office Clipboard as a panel in each application.

Exercise 2: Object Embedding

To get the 'flavour' of OLE carry out the following:

(a) In the CHAP15.XLS workbook click once on the chart. Click the Copy button.

(b) Move to your word processor and start a new document. Click the Paste button to copy the chart.

(c) Double click the chart. If you are new to OLE the result is unexpected. Although you are running a word processor application (Word, WordPerfect, etc.), the part of the screen containing the chart now looks like Excel. That is exactly what it is. The whole of your Excel workbook has been *embedded* in the document. While the chart is open, the application's menus and tools bars have been replace by those from Excel.

Your embedded workbook should consist of a chart sheet and the sheets that were present in the original workbook. Remember to go back to the chart sheet before closing the embedded object since this is what you want displayed.

(d) If you save the word processing document, a copy of the workbook is saved with it; not as a separate file but as part of the word processing document file. You could give a copy of the file to a colleague and he/she could modify the workbook provided his/her computer had Excel installed.

Embedding vs Linking

In the previous exercise we have *embedded* a workbook in a document. Part of the word processing 'space' has Excel properties. In the embedded object of Exercise 2 we displayed a chart. We can open the workbook from within the word processing application, modify the data and hence update the chart.

Is *linking* the same? Yes and no! If you link a workbook to a document, the workbook is accessible from it provided the workbook file is present in the same folder that it was when the linking was made. One way to think of linking is to imagine that what you see in the document is a picture of the part of the workbook to which it is linked.

Consider the following scenario.

1. An Excel workbook is linked to a Word document on Monday. Clearly, the workbook and the document display the same data.

2. On Tuesday, the data in the workbook is revised. Since the document file is not open, its data is now out of date.

3. On Wednesday, when the document is opened, the word processor will display a message stating that the document contains links and asking if you wish to update them now. If you reply Yes, the workbook is opened but you do not see this happen. The data is updated and the workbook is closed.

If nothing was done to the workbook on Tuesday, the same message would be displayed on Wednesday since the system has no way of knowing if it has been revised since the last time the document was used.

Linking has certain advantages over embedding: (1) it is often easier to revise the workbook by opening it on its own, (2) no matter what part of the workbook is active, the document displays the same data it did when the link was first established, and (3) the document file size is not as large as with embedding.

Exercise 3: Embedding and Linking

This is a do-it-yourself exercise. We are near the end of the book and by now you do not need to be told every step. The task is to embed and to link the chart in CHAP15.XLS with a word processor document and to experiment with the results.

But I have not told you how to link! Look at Figure 15.3 and note the two radio buttons. *Paste* results in OLE *embedding*, *Link* results in OLE *linking*. In the *As* box select either Microsoft Excel Worksheet or Microsoft Excel Chart.

It is also possible to use OLE within a word processing document with the Insert|Object... command. You may wish to experiment with this.

Exercise 4: Creating an Equation

$$\int_0^{\pi} \frac{1}{x^2} dx$$

Applet: An applet is a small application which must be run from within another application.

Microsoft provides an applet called Equation Editor which may be used in programs such as Word or Excel to create an equation. With a little practice and experimentation you will be able to create complex equations. In this exercise we create the expression at the left to get you started.

(a) On Sheet2 of CHAP15.XLS, use the command Insert|Object... and select *Microsoft Equation 3.0*. Figure 15.4 shows how the worksheet appears.

(b) To draw the integral sign, click the mouse pointer over the fifth item on the bottom row of the Equation Editor toolbar. Move the pointer to the second item on the top row of the drop down menu since we need an integral sign with two limits.

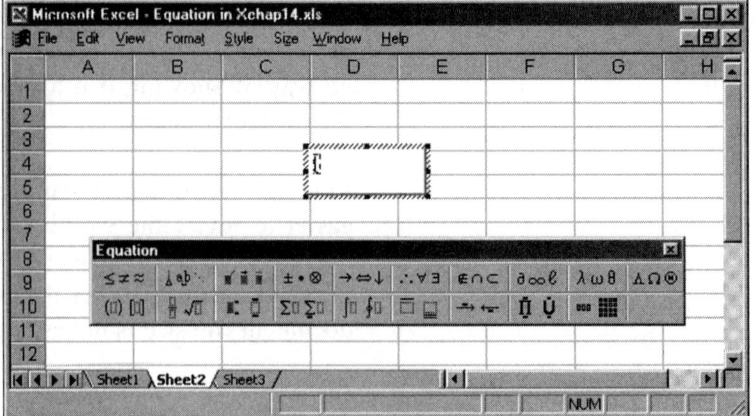

Figure 15.4

(c) Experiment by tapping the [Tab] key; hold down the [Shift] key and tap [Tab]. The L shape that moves around is the *insertion point*. When a box has something typed in it, the L is reversed. Now use the mouse to move the insertion point to the box which will hold the lower limit. In this box type 0.

(d) Using either the mouse or [Tab], move to the box where the upper limit will go. Open the ninth item on the top row of the toolbar and click on the π symbol.

(e) Use [Tab] to move the insertion point into the box at the right of the integral sign. Move the pointer to the second item on the bottom row of the toolbar and select the first item on the top row of the drop down menu – two open boxes stacked vertically with a bar between them. We need this template for the $1/x^2$ part of the expression.

(f) Move the insertion point to the top box and type 1. Move the insertion point to the bottom box and type x. Experiment using the mouse to relocate the insertion point.

(g) We need an object to hold the superscript. Move to the third item on the bottom row of the toolbar and select the first item from the menu. You should now have a superscript box in which to type the 2. Alternatively, type the 2, select it, and now open the template for a superscript.

(h) Press [Tab] to move the insertion point to the far right of the equation. Type dx.

(i) Click the mouse anywhere outside the equation box to close the Equation Editor applet.

As you may have discovered, you cannot use [Spacebar] when forming an equation – the applet looks after the spacing of items. You can however, use [Ctrl]+[⇧ Shift] to add addition spacing.

By default the Equation Editor uses italics for variables such as *x* and regular font for digits and anything it thinks is a function such as Exp or Ln. You can enter normal text, including spaces, in an equation box by using the Text item in the Style menu.

Some users have reported that they have better success if they compose the equation in Word and copy the completed object to the Excel worksheet.

Exercise 5: Interactive Web Page

The techniques (Copy and Paste, or Copy and Paste Special) we used in Exercise 1 may be used to place data and pictures of charts on web pages. Some simple experimenting will quickly show you the correct method.

Microsoft Excel 2000 and Internet Explorer 4 introduced a new concept: web pages with interactive Excel. We will briefly explore this topic but you will need Microsoft FrontPage or a similar web page composer to develop more complex pages. The HTML files made in the exercise will be saved on the local hard drive. You will, of course, need to move them to your web folder if you wish Internet users to be able to access them.

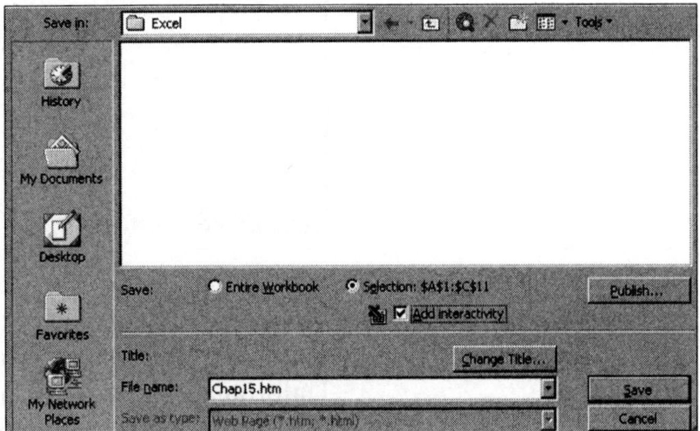

Figure 15.5

(a) Open Sheet1 of CHAP15.XLS and select the range A1:D11. Open the File menu and select *Save as Web Page*. This opens up a dialog box similar to Figure 15.5. If you merely click the *Save* button, a non-interactive web page will be made.

‡ When a web page has previously been saved from an Excel workbook, this button may be called *Republish*.

(b) Click in the *Selection* radio button‡ and in the *Add interactivity* option box. Enter Chap15.htm in the *Name* box. Click the *Save* button.

(c) Open the Chap15.htm file with Internet Explorer (version 4 or better). The page will be similar to that in Figure 15.6. Experiment by changing some of the values in the table and noting the change in the value in B11.

You may wish to ensure that the users can change only certain cells. Before saving the Excel file as a web page, select the cells that the user may change and use Format|Cells to unlock them. Then protect the worksheet using Tools|Protection.

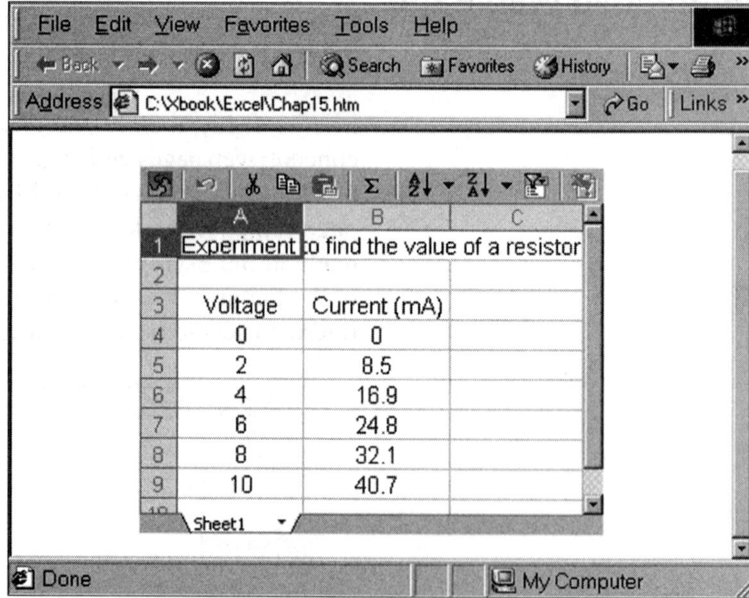

Figure 15.6

You may ask why the chart was not included in step (a). Unfortunately, even if it had been included in the selection, the chart would not have appeared on the web page. Putting a chart on the web page requires a little more work.

(d) With the chart selected, open the File menu and click *Save as Web Page*. Select the *Selection* (or *Republish*) radio button and put a check mark in the *Add interactivity* box. Use the name Chap15b.htm and click the *Save* button.

(e) When you open the new file with Internet Explorer the chart and part of the worksheet are visible. However, it is disappointing to find that only the cell A3:B9 (i.e. the ones used to make the chart) have data in them. The chart is interactive but we get no value in B11.

The process in step (d) gave us a web page with a chart but only those rows of the worksheet needed by the chart are displayed. So we will cheat! We will make the chart depend on more cells. We need the chart to be dependent on data in row 1 and row 12.

(f) Move to the worksheet and right click the chart. Select *Source Data* and open the *Series* tab. Add a new series with *x*-values as E1 and *y*-values as F1. Add another new series with A12 as the *x*-values and B12 as the *y*-values. Since there is no data in these cells, the chart is unchanged. But as far as Excel is concerned the chart is dependent on them.

(g) Repeat step (d) and save the web page as Chap15c.htm. Open the new file with Internet Explorer. Now we have both a chart and a worksheet that includes all the cells of interest.

To use Microsoft Excel interactive data on the Web, your users must have Microsoft Office XP or access to an Office XP licence and the Office Web Components installed. Furthermore, there are some limitations you should learn about before developing a large web page. In Excel Help search with the word *Guidelines* and open the topic *Guidelines and limitations for saving or publishing Web pages*. Strangely, this topic does not show when the phrase *web page* is used for the search!

You may also wish to experiment with using the *Publish* button in place of *Save*. This offers some options for the web page. It will generate both an HTML file and a similarly named folder which hold files needed for the web page.

Answers to Starred Problems

Chapter 2

1. (a) =2*A1 – C1
   (b) =B1^2 + A1
   (c) =1/(C1^2 – D1^2) or =(C1^2 – D1^2)^–1
   (d) = (A1 + B1) / (C1 – D1)

2. (a) =D1^0.5 or =D1^(1/2)
   (b) =D2^(1/3)
   (c) =D3^(–1) or =1/D3 (the latter is better)

Chapter 4

1. =ROUNDUP(Length * Height / 2.25)

3 MINVERSE is an *array formula*. You should have selected F3:I6, typed the formula and then pressed Ctrl+Shift+Enter.

4. =DEGREES(ASIN(OPP/HYPOT))

5. =FACT(A10) / FACT(A10 – B10) or =PERMUT(A10, B10)

Chapter 5

1. =IF(A1=0, 1, SIN(X)/X) or =IF(A1<>0, SIN(X)/X, 1)

2. =IF(x<2, IF(y>=10, IF(y<=12, IF(z>=0,"Pass","Fail"), "Fail"), "Fail"), "Fail")

3. =IF(AND(x<2, AND(y>=10, y<=12), z>=0), "Pass", "Fail")

4. There are many possible solutions. The simplest is to use:
   D2: =IF(A2+B2=2,1,0), E2: =IF(A2+B2>=1,1,0)
   F2: =IF(C2+E2=2,1,0), G2: =IF(D2+F2>=1,1,0)

   Replacing the 0s and 1s by FALSE and TRUE values in the A, B and C columns, we could use either of these formulas:
   D2: =AND(A2, B2), E2: =OR(A2,B2)
   F2: =AND(C2, D2), G2: =OR(D2,F2)

   or, with greater risk of making an error, we may use just the D column with D2 having the formula:
   =OR(AND(A2,B2),AND(C2,OR(A2,B2)))

5. Name the range A3:F15 as ColourCode and enter these formulas:
   A21: =VLOOKUP(A19,ColourCode,2, FALSE)
   B21: =VLOOKUP(B19,ColourCode,2, FALSE)
   C21: =VLOOKUP(C19,ColourCode,2, FALSE)
   D21: =VLOOKUP(D19,ColourCode,2, FALSE)
   A22: =VLOOKUP(A19,ColourCode,3, FALSE)
   B22: =VLOOKUP(B19,ColourCode,4, FALSE)
   C22: =VLOOKUP(C19,ColourCode,5, FALSE)
   B23: =(A22*10+B22)*C22
   C23: =VLOOKUP(D19,ColourCode,6, FALSE)

Chapter 6

2. After opening the *Source Data* dialog, move to the *Series* tab. Click the *Add* button and proceed to add the new data. Remember you must specify both the y-values and the x-values.

3. Make the chart in the normal manner. In an unused part of the worksheet enter a pair of x- and y-values in two adjacent cells. Values such as 3 and 0.9 will do. Add these to the chart as a new data series. The Copy|Paste Special method explored in Exercise 10 may be use. Alternatively, right click on the chart and use the Add button on Source Data dialog.

   Right click on the new point and open the *Format Data Series* dialog. On the *Axis* tab specify *secondary axis.* Return to the chart and right click. On the *Axes* tab of the *Chart Options* dialog, put a ✓in the *Value (X) axis* box. Format the two secondary axes such that they have the same minimum, maximum and units as their respective primary axis. When all is ready, format the new data series to have no line and no markers – to be invisible.

Chapter 7

5. The results are (a) for length vs length: $b = 1.0275$, $R^2 = 0.988$, and (b) for area vs length: $b = 1.8065$, $R^2 = 0.9931$. You could plot Y against X and insert a trendline for a Power model rather than a Linear model. If you try to make a log–log plot of the data in the table, a message pops up stating *Negative or zero values cannot be plotted correctly on log charts*. To plot a log–log chart, first multiply the lengths by 10 to give values in millimetres and the area values by 100 to give millimetres squared.

Chapter 8

1. The required functions may be coded as shown below.

```
Function Kelvin(Fahrn)
    Celsius = (Fahrn - 32) * 5 / 9
    Kelvin = Celsius + 273.15
End Function
```

Of course, the two statements could be combined: Kelvin = (Fahrn-32)*5/9 + 273.15

```
Function Stirling(n)
    Stirling = n * Log(n) - n
End Function

Function SumRange(a, b)
    If a <= b Then
        lower = a:   upper = b
    Else
        lower = b:   upper = a
    End If
    SumRange = 0
    For n = lower To upper
        SumRange = SumRange + n
    Next n
End Function
```

Note the use of the colons in *SumRange* to place two statements on one line. The formula =LN(FUNC(A3)) may be used to compare the Stirling approximation with the exact value. When A3 is 100, the approximation gives 360.5 while the exact value is 363.7. One cannot test values of *n* above 170 since 170! is close to Excel's maximum value of 1E+307.

2.

```
Option Base 1
Function Quad2(a, b, c)
    Dim Temp(3, 1)
    d = (b * b) - (4 * a * c)
    Select Case d
     Case Is < 0
      Temp(1, 1) = "No real"
        Temp(2, 1) = "roots": Temp(3, 1) = ""
     Case 0
        Temp(1, 1) = "One root"
        Temp(2, 1) = –b / (2 * a): Temp(3, 1) = ""
     Case Else
        Temp(1, 1) = "Two roots"
        Temp(2, 1) = (-b + Sqr(d)) / (2 * a)
        Temp(3, 1) =  (-b - Sqr(d)) / (2 * a)
    End Select
    Quad2 = Temp
End Function
```

Chapter 9

1(a)By the principle of mass balance, if V ml of water is added to V_a ml of solution of concentration C_a, and the concentration of the resulting solution is C_b, then $C_aV_a = C_b(V + V_a)$. In this problem V_a is always 1, so we may write $C_b = C_a/(1 + V)$.

The required formulas are:
B11: =IF(B10>req, B10/(vol+1), NA())
C11: =IF(B10>req, C10+(wash+ vol*disp/1000), NA())

1(b)From the equation developed in (a), $C_1 = C_0/(V + 1)$ and $C_2 = C_1/(V + 1)$. By substitution, we may write the last equation as $C_2 = C_0/(V + 1)^2$. Continuing with this, we get $C_n = C_0/(V + 1)^n$. Take the logarithm of both sides to find what value of n gives the required final concentration. The formulas are:
B9: =CEILING(LN(init/req)/LN(A9+1),1)
C9: =(B9*wash + B9*A9/1000*disp)

We need an integer value in B9 but the INT function is not appropriate; if the formula evaluates to 4.23 we need a value of 5 not 4 to ensure that the concentration is equal to or less than the required value. The CEILING or ROUNDUP function is needed.

2. The equation for the decay of iodine-135 integrates to $[\mathrm{I}]_t=[\mathrm{I}]_0 \exp(-k_I t)$. The series expansion of $\exp(-x)$ is $1 - x + x^2/2! - x^3/3!\ldots$ Hence, provided $-k_I t$ is small (i.e. small time intervals and reasonably large half-life), our approximate should give values of acceptable accuracy. Your graph should resemble that below.

The exact solution for xenon production is:

$$[\mathrm{Xe}]_t = [\mathrm{Xe}]_0 \exp(-k_{\mathrm{Xe}}t) + \frac{k_{\mathrm{I}}[\mathrm{I}]_0}{(k_{\mathrm{I}} - k_{\mathrm{Xe}})}(\exp(-k_{\mathrm{Xe}}t) - \exp(-k_{\mathrm{I}}t))$$

You may wish to add these two exact solutions to your worksheet and the chart. The agreement is most satisfactory. Clearly, when the exact solution to a problem is known, it should be used rather than an approximation. However, when a carefully developed approximation agrees with an exact solution, one has some added confidence that an error did not slip into the derivation of the exact solution. The converse is not necessarily true.

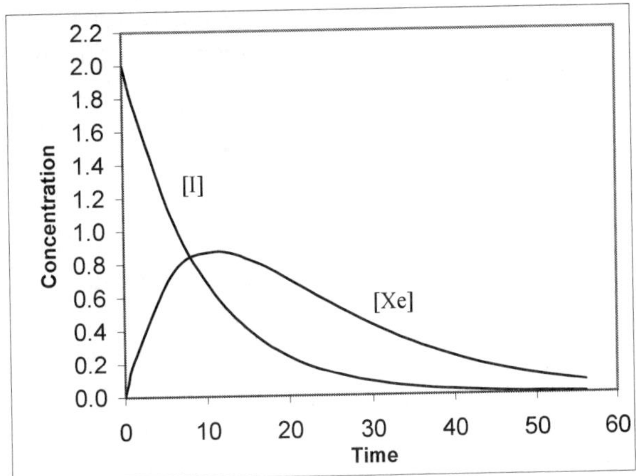

3. The required formulas are shown in the figure below. The *Quick* values are very close to the *Sine* values.

	A	B	C	D	E	F
1	n	d	q	Quick	Angle	Sine
2	1	-1	128	=C2/128	90	=SIN(RADIANS(E2))
3	=A2+1	=IF(C2>=0,B2-1,B2+1)	=C2 + B3	=C3/128	=E2 - 5.625	=SIN(RADIANS(E3))

Chapter 10

3. This problem needs a precision of at least 1E−09. The table below shows the formulas and constraints needed to run Solver. The results are somewhat dependent on the starting values. These were obtained with starting values of 2 for *N* and *V*. Adding a constraint to make *V* an integer seems to overtax Solver.

Solver Solution		**Formulas**	**Constraints**
V	84.1		V > 0
N	4		N > 1, N = int
conc	9.55E–09	=init*(1+V)^–N	conc<=1E–08
cost	1.50	=N*wash + N*V*disp/1000	

Chapter 11

2. With an extreme case such as $p = 1$, $q = 10$, look at the values in each strip. Essentially, only the first strip makes a significant contribution to the total. A plot of *x* against *y* is also instructive.

3. The data for this problem was generated using the function $\ln(x^2 - 4)$. The exact solution for this is given by

$$\int_1^2 \ln(x^2 - a^2)dx = x\ln(x^2 + a^2) - 2x + 2a\tan\left(\frac{x}{a}\right)^2_1$$

From the data below, clearly Simpson's rule gives the most accurate result.

Trapezoid	1.744246	single strips
Trapezoid	1.652087	pairs of strips
Romberg	1.774966	
Simpson ⅓	1.836447	
Exact	1.836447	

4. Simpson's rule cannot be used since the strips are not equally spaced. The trapezoid rule gives a value of 9.792705. If there is reason to believe the data should fit a smooth curve, plot the data and find the trendline equation. The data in this question fits $y=0.080716x^2-0.936796x+3.301219$ with $R^2=0.99734$. Integration of this equation gives 9.744195.

6. You need to discover that the two curves intersect at (0,0) and (2,16). Simpson's rule could be applied to the separate curves and the results subtracted. A better way is the divided the region R into strips of width dx and height $4x^2 - x^4$.

Chapter 12

1. The exact solution of $y' = -xy^2$ is $y = 2/(1 + x^2)$.

Chapter 13

1. With large c values, the motion decays exponentially. With small values, the amplitude of the harmonic motion decreases with time.

A possible worksheet for the Van der Pol's equation is shown below together with the resulting graphs. The formulas are:

```
B8:  =x0
C8:  =v0
D8:  =h*van(B8, C8)
E8:  =h*van(B8+h*C8/2, C8+D8/2)
F8:  =h*van(B8+h*C8/2+h*D8/4, C8+E8/2)
G8:  =h*van(B8+h*C8+h*E8/2, C8+F8)
B9:  =B8+h*C8+(h/6)*(D8+E8+F8)
C9:  =C8+(1/6)*(D8+2*E8+2*F8+G8)
```

D8:G8 is copied to row 9; row 9 is copied down as far as required.

The user-defined function *van* is coded as:

```
Function van(x, v)
   c = 2
   van = –c * (x ^ 2 –– 1) * v –– x
End Function
```

	A	B	C	D	E	F	G
1	Van der Pol's Equation			dv/dt + c(x^2 -1) +x = 0			
2							
3	initial x	x0	0.75				
4	initial v	v0	0				
5	increment	h	0.05				
6							
7	time	displace	velocity	k1	k2	k3	k4
8	0.00	0.750	0.000	-0.0375	-0.0383	-0.0383	-0.0391
9	0.05	0.749	-0.038	-0.0391	-0.0400	-0.0400	-0.0408
10	0.10	0.746	-0.078	-0.0408	-0.0416	-0.0416	-0.0425

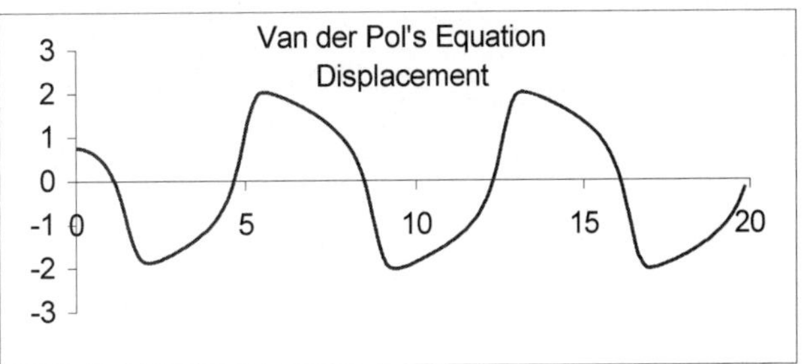

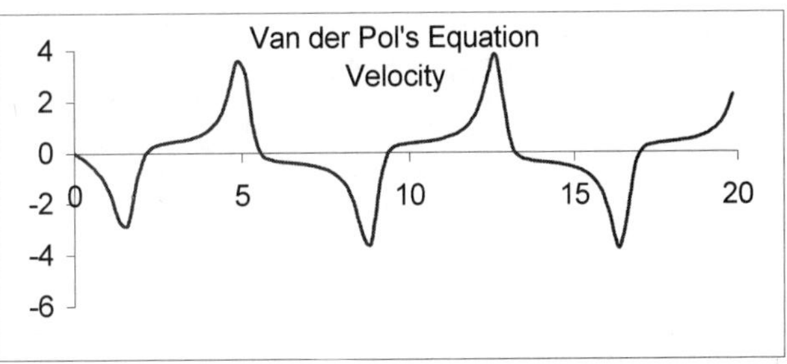

2. The required worksheet is shown in the figure below. The rows are repeated down as far as required; say that is row 30. Not shown are cell names *k*, *c*, *p*, *L*, *intervals*, *deltaX* and *deltaT*. The B and J columns reflect the boundary conditions $u(0,t) = 0$ and $u(2,t) = 0$. The top row (B3:J3) reflects the initial

condition – in this problem use =100Sin(PI()*B3/L) in B4 and copy it to B4:J4. The formula in C5 is =(B4 + D4)/2 and copy it to C5:F30. Now we need to work from the other end. In I5 enter =(H4 + J4)/2 and copy this to I5:G30. Note that the data in the columns for $x = 0.25$ and $x = 1.75$ is as expected. The exact solution is $u = 100\exp(-\pi^2 kt/cpL^3)\sin(\pi x/L)$.

	A	B	C	D	E	F	G	H	I	J
1	Heat Flow in Plate									
2						x values				
3	t	0	0.25	0.5	0.75	1	1.25	1.5	1.75	2
4	0.00000	0	38.27	70.71	92.39	100.00	92.39	70.71	38.27	0
5	0.20625	0	35.36	65.33	85.36	92.39	85.36	65.33	35.36	0
6	0.41250	0	32.66	60.36	78.86	85.36	78.86	60.36	32.66	0

You should now try to model the situation in which the entire plate is initially at 100°C and one face is suddenly raised to 200°C.

Index